Computational Techniques for Biological Sequence Analysis

This book provides an overview of basic and advanced computational techniques for analyzing and understanding protein, RNA, and DNA sequences. It covers effective computing techniques for DNA and protein classifications, evolutionary and sequence information analysis, evolutionary algorithms, and ensemble algorithms. Furthermore, the book reviews the role of machine learning techniques, artificial intelligence, ensemble learning, and sequence-based features in predicting post-translational modifications in proteins, DNA methylation, and mRNA methylation, along with their functional implications. The book also discusses the prediction of protein–protein and protein–DNA interactions, protein structure, and function using computational methods. It also presents techniques for quantitative analysis of protein–DNA interactions and protein methylation and their involvement in gene regulation. Additionally, the use of nature-inspired algorithms to gain insights into gene regulatory mechanisms and metabolic pathways in human diseases is explored. This book acts as a useful reference for bioinformaticians and computational biologists working in the fields of molecular biology, genomics, and bioinformatics.

Key Features:

- Reviews machine learning techniques for DNA sequence classification and protein structure prediction
- Discusses genetic algorithms for analyzing multiple sequence alignments and predicting protein–protein interaction sites
- Explores computational methods for quantitative analysis of DNA–protein interactions
- Examine the role of nature-inspired algorithms in understanding the gene regulation and metabolic pathways
- Covers evolutionary algorithms and sequence-based features in predicting post-translational modifications

Computational Techniques for Biological Sequence Analysis

Edited by
Saiyed Umer
Ranjeet Kumar Rout
Monika Khandelwal
Smitarani Pati

CRC Press
Taylor & Francis Group
Boca Raton London New York

CRC Press is an imprint of the
Taylor & Francis Group, an **informa** business

Designed cover image: Shutterstock

First edition published 2025
by CRC Press
2385 NW Executive Center Drive, Suite 320, Boca Raton FL 33431

and by CRC Press
4 Park Square, Milton Park, Abingdon, Oxon, OX14 4RN
CRC Press is an imprint of Taylor & Francis Group, LLC

ISBN: 978-1-032-63026-7 (hbk)
ISBN: 978-1-032-66070-7 (pbk)
ISBN: 978-1-032-66071-4 (ebk)

DOI: 10.1201/9781032660714

Typeset in Nimbus font
by KnowledgeWorks Global Ltd.

Contents

About the Editors

Saiyed Umer is currently serving as an Assistant Professor in the Department of Computer Science and Engineering, Aliah University, Kolkata, India. He was the Research Personnel at Indian Statistical Institute (ISI), Kolkata, India, from November 2012 to April 2017. He received a Ph.D. degree (Engineering, executed in ISI Kolkata) from the Department of Information Technology at Jadavpur University, Kolkata, India, in March 2017. He earned a B.Sc. (Hons) degree in Mathematics from Vidyasagar University, India, in 2005, and a Master of Computer Applications from the West Bengal University of Technology, India, in 2008. Dr Umer received an M.Tech degree from the University of Kalyani, India, in 2012. He has published several papers in peer-reviewed international and scientific journals in the fields of biometrics, affective computing, big-data research, business human resource management, and computational biology. His research interests include computer vision, machine learning, deep learning, and business data analytics techniques.

Ranjeet Kumar Rout is currently serving as Assistant Professor in the Department of Information Technology, Dr. B. R. Ambedkar National Institute of Technology, Jalandhar, Punjab, India. Formally, he was the Assistant Professor in the Department of Computer Science and Engineering, National Institute of Technology Srinagar, Hazratbal, India. He received his Ph.D. from the Department of Information Technology of Indian Institute of Engineering Science and Technology, Shibpur, West Bengal, India. Previously, he earned post graduate and bachelor's degrees in Computer Science and Engineering from Biju Patnaik University of Technology, Odisha, India, in 2010 and 2005, respectively. Prior to working at NIT Srinagar, Dr. Ranjeet had research and teaching experience from Amity University Noida, National Institute Technology Jalandhar, and Indian Statistical Institute (ISI), Kolkata, India. His research interests include machine learning, deep learning, visual cryptography, and computational biology. He has published several papers in peer-reviewed international and scientific journals in the field of non-linear Boolean functions and computational biology.

Monika Khandelwal is currently an Assistant Professor in the Department of Computer Science and Engineering, Jaypee University, Solan, Himachal Pradesh, India. She earned her Ph.D. from the Department of Computer Science and Engineering at National Institute of Technology Srinagar, Hazratbal, India. She received her M.Tech. degree in Computer Science and Engineering from Dr. B. R. Ambedkar National Institute of Technology Jalandhar, Punjab, India, in 2016 and her B.Tech. degree in Computer Science and Engineering from Guru Jambheshwar University of Science & Technology, Hisar, Haryana, India. Prior to obtaining her Ph.D. at NIT Srinagar, she had teaching experience from National Institute Technology Hamirpur and Malaviya National Institute of Technology Jaipur, India. She has published several papers in conferences, journals, and book chapters. Her research interests include machine learning, deep learning, bioinformatics, and computational biology.

Smitarani Pati is working in Instrumentation and Control Engineering on modeling, control, and optimization of industrial processes such as energy optimization using soft computing techniques. She earned her Ph.D. degree in Instrumentation and Control Engineering from Dr. B. R. Ambedkar National Institute of Technology, Jalandhar, Punjab, India. She received her B. Tech. degree in Electrical Engineering from the Biju Patnaik University of Technology, Odisha, India, in 2011, and her M.Tech. degree in Control and Instrumentation Engineering from Dr. B. R. Ambedkar National Institute of Technology, Jalandhar, Punjab, India, in 2018. She has published several articles in international conferences and book chapters. Her current research interests include energy modeling and optimization, design of distributed systems, and fault-tolerant controls.

Contributors

Parul Agarwal
CSE and IT Department
Jaypee Institute of Information Technology
Noida, Uttar Pradesh, India

Navya Aggarwal
Centre for Computational Biology and Bioinformatics
Amity Institute of Biotechnology
Amity University
Noida, Uttar Pradesh, India

Shaik Akbar
Department of Computer Science
Potti Sriramulu Chalavadi Mallikarjuna Rao College of Engineering and Technology
Vijayawada, Andhra Pradesh, India

Hina Bansal
Centre for Computational Biology and Bioinformatics
Amity Institute of Biotechnology
Amity University
Noida, Uttar Pradesh, India

Savvy Bansal
CSE and IT Department
Jaypee Institute of Information Technology
Noida, Uttar Pradesh, India

Mayank Roy Chowdhury
Department of Biotechnology
National Institute of Technology
Tadepalligudem, Andhra Pradesh, India

Renu Dhir
School of Computer Science and Engineering
Lovely Professional University
Phagwara, Punjab, India

Sana Farooq
Government College of Engineering and Technology
Safapora, Jammu & Kashmir, India

Kamalpreet Kaur
School of Computer Science and Engineering
Lovely Professional University
Phagwara, Punjab, India

Jewaira Khurshid
Government College of Engineering and Technology
Safapora, Jammu & Kashmir, India

Sanjeev Kumar
Biomedical Instrumentation Unit
CSIR—Central Scientific Instruments Organisation (CSIR-CSIO)
Chandigarh, Punjab, India

Kabita Kumari
Department of Instrumentation and Control Engineering
Dr. B. R. Ambedkar National Institute of Technology
Jalandhar, Punjab, India

Richa Kumari
Centre for Computational Biology and Bioinformatics
Amity Institute of Biotechnology
Amity University
Noida, Uttar Pradesh, India

Ruchi Jakhmola Mani
Centre for Computational Biology and Bioinformatics
Amity Institute of Biotechnology
Amity University
Noida, Uttar Pradesh, India

Shikha Mehta
CSE and IT Department
Jaypee Institute of Information Technology
Noida, Uttar Pradesh, India

Amogh Misra
CSE and IT Department
Jaypee Institute of Information Technology
Noida, Uttar Pradesh, India

Ishrat Nazeer
Department of CSE
National Institute of Technology
Srinagar, Jammu & Kashmir, India

Sharvan Kumar Pahuja
Department of Instrumentation and Control Engineering
Dr. B. R. Ambedkar National Institute of Technology
Jalandhar, Punjab, India

Praveen Pathak
CSE and IT Department
Jaypee Institute of Information Technology
Noida, Uttar Pradesh, India

S.G.S. Priyanka
Department of Computer Science
Potti Sriramulu Chalavadi Mallikarjuna Rao College of Engineering and Technology
Vijayawada, Andhra Pradesh, India

A. Pathanjali Sastri
Department of Computer Science
Potti Sriramulu Chalavadi Mallikarjuna Rao College of Engineering and Technology
Vijayawada, Andhra Pradesh, India

K. Suresh
Department of Computer Science
Potti Sriramulu Chalavadi Mallikarjuna Rao College of Engineering and Technology
Vijayawada, Andhra Pradesh, India

Kunwar Vidhan
CSE and IT Department
Jaypee Institute of Information Technology
Noida, Uttar Pradesh, India

Sudarshana Deepa Vijaykumar
Department of Biotechnology
National Institute of Technology
Tadepalligudem, Andhra Pradesh, India

1 Machine Learning and Computational Models for the Prediction of Post-Translational Modification Sites

Richa Kumari, Navya Aggarwal, Ruchi Jakhmola Mani, and Hina Bansal

1.1 INTRODUCTION TO POST-TRANSLATIONAL MODIFICATIONS (PTMS)

Post-translational modifications (PTMs) influence the diverse functionality, of proteins which is important in cellular biology. After the production of protein, these specific chemical changes control the position, shape, stability, and activity of proteins, performing as molecular switches in response to biological stimuli [1]. PTMs act with respect to cells to precisely control activities of protein in different biological signals. The expression PTMs refers to a series of possible chemical changes that can occur in the amino acid residues of the protein. They can undergo many processes including nitrosylation, ubiquitination, phosphorylation, glycosylation, acetylation, methylation,SUMOylation, ADP-ribosylation, lipidation and many more. Each of these PTMs modifies the structure of the protein or adds new chemical groups, giving it specific regulatory properties [2, 3]. Recent study evidence shows that human proteomes are significantly more complex than genomes. The human genome contains between 20,000 and 25,000 genes; meanwhile, the human proteome is speculated to contain around a million different proteins [2, 4]. These data show that several numbers of proteins are generated by one gene. Numerous processes, including recombination of B chromosomes, alternative transcription start sites, differences in transcription termination, and alternative splicing of B transcripts influence mRNA protein diversity. PTMs greatly aid the difficult process of converting DNA into proteome [5].

DOI: 10.1201/9781032660714-1

1.1.1 SIGNIFICANCE OF POST-TRANSLATIONAL MODIFICATION SITE

PTMs play a classified and important role as regulators in various biological processes and functions, which can be explained as follows:

1. **Transduction of signals**: Phosphorylation, one of the most well-known PTMs, controls cell signaling pathways. It allows cells to respond to external stimuli by changing the state of proteins in various signalling pathways [6].
2. **Gene Expression**: Chemical revision, such as the insertion or deletion of methyl or acetyl groups, can alter the chromatin structure that controls gene expression after proteins are created. Cell and development hold the epigenetic modifications, which are one of the classes of post-macromolecular protein alterations.
3. **Protein Stability and Degradation**: The process of acetylation and ubiquitination are two instances of PTMs that significantly impact protein stability and proper folding [7]. These modifications have an impact on cells' general proteostasis, which maintains healthy and functional cells by maintaining stable and correctly folded proteins.
4. **Cellular Localization**: Cellular localization is a complex process mostly attributed to PTMs-mediated spatial regulation. A cell must be carefully localized to its specific subcellular compartments to interact with the right partners and participate in the right molecular signaling pathways, which is required for optimal cellular function. PTMs play an important role in controlling this spatial coordination and ensure that each cellular component should work in a coordinated and orderly manner. Further, PTMs are an essential component of the spatial control of targeting proteins to specific cell field, in resulting the accurate and precise localization and interaction of proteins [8,9].
5. **Enzyme Activation**: Enzyme catalytic activities can be altered by PTMs. It can increase or decrease the activity of enzymes by activating or deactivating them [10]. For example, kinases can be activated by phosphorylation and thus participate in various biological activities.

1.2 CATEGORIES OF SITES FOR POST-TRANSLATIONAL MODIFICATION

PTMs can be categorized in various categories:

1.2.1 PHOSPHORYLATION

Phosphorylation is a process in which a phosphate group binds to an amino acid residues like tyrosine, threonine, and serine [11]. Enzymes-phosphatases, and

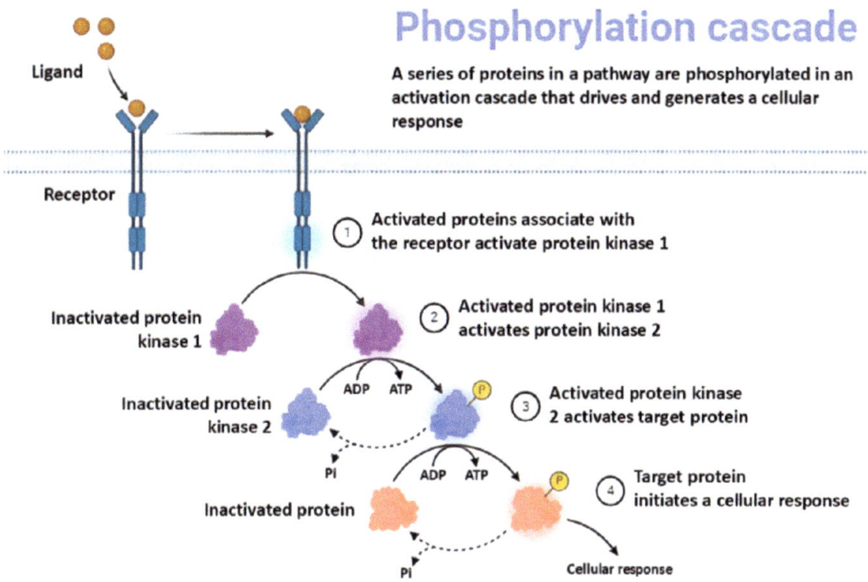

FIGURE 1.1 The cascade of phosphorylation in proteins. (Created with `BioRender.com`.)

kinases, which act as molecular switches in the signaling pathways of many cells, are responsible for this change. Phosphorylation plays a critical role in immune system responses, cell cycle regulation, and cell signaling, among others [12]. This has important consequences for the function, localization, and interaction of proteins in the cell. Figure 1.1 illustrates an example of the phosphorylation cellular response. It represents how a series of proteins are phosphorylated in an activated environment of protein kinase that generates a cellular response.

1.2.2 GLYCOSYLATION

The process of attaching sugar molecules (glycans) to proteins, called glycosylation, has significant effects on protein structure and function. N- and O-linked glycosylation are the two main variants of this alteration [11]. The mechanisms of protein folding, stability, and cellular recognition largely depend on glycosylation. Glycan analysis requires the use of specialized techniques, including mass spectrometry and lectin-based assays [13]. Figure 1.2 illustrates an example of the process of glycosylation on protein in which N-glycans are attached to the asparagine residues of protein.

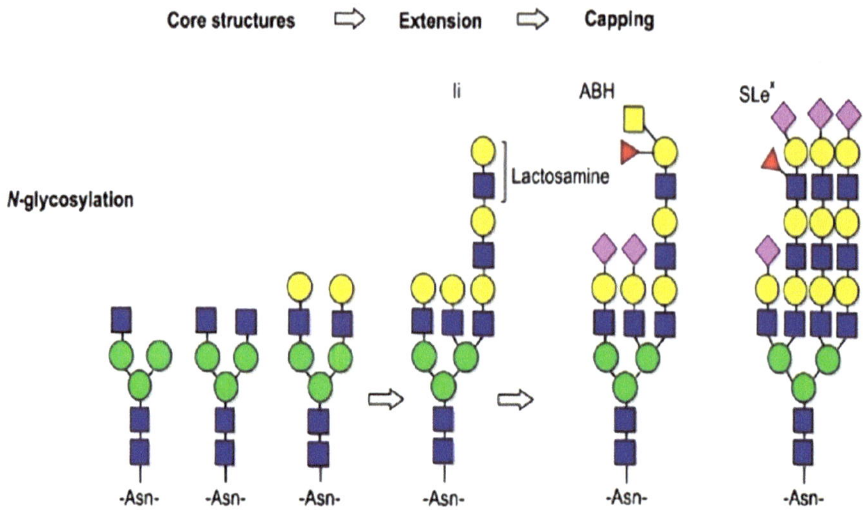

FIGURE 1.2 The glycosylation on protein (N-glycans are attached to the asparagine residues of protein). (Created with `BioRender.com`.)

1.2.3 ACETYLATION AND METHYLATION

Modifications after translation, such as acetylation and methylation, are frequently observed on lysine and arginine residues of histone proteins [14–16]. Chromatin structure and gene expression are determined by these changes. Depending on the environment, methylation can have activating or repressive effects, but acetylation is generally linked to gene activation [17]. These PTMs are essential for important biological functions such as chromatin remodeling and transcriptional control [2] (Figure 1.3).

1.2.4 UBIQUITINATION AND SUMOylation

The process of attaching compact protein tags such as ubiquitin or SUMO to specific target proteins is called ubiquitination or SUMOylation [18]. Ubiquitination controls the balance of protein turnover and primarily targets proteins to be destroyed via the proteasome. On the other hand, SUMOylation controls how proteins interact, where cells are located, and how transcription works [19, 20]. For cell signaling and homeostasis to be maintained, these two changes are essential (Figure 1.4).

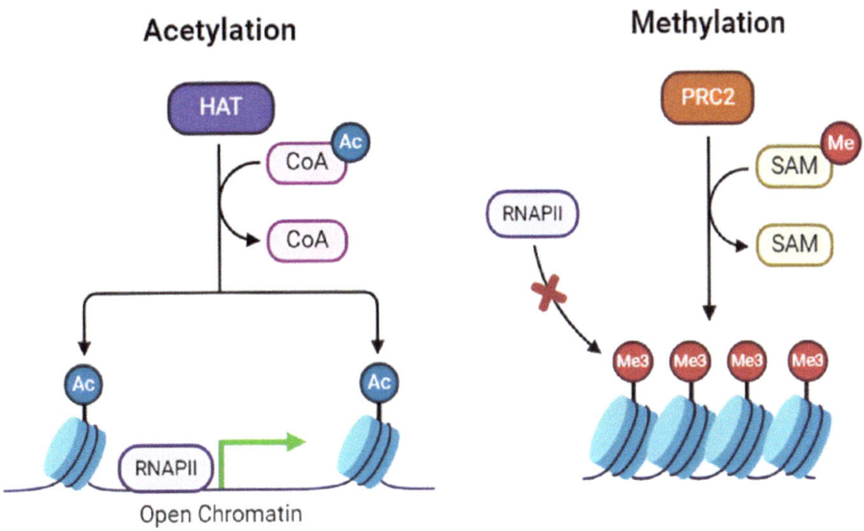

FIGURE 1.3 Complex group incorporation: Acetylation and methylation. (Created with `BioRender.com`.)

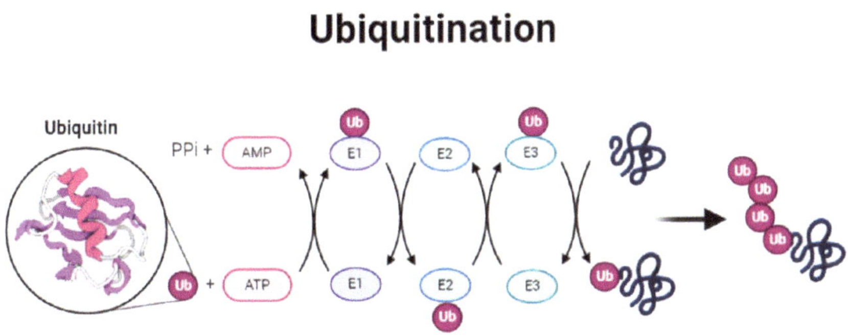

FIGURE 1.4 Similar cycle between SUMOylation and ubiquitination. (Created with `BioRender.com`.)

1.2.5 ADDITIONAL POST-TRANSLATIONAL MODIFICATIONS (PTMS): NITROSYLATION, ADP-RIBOSYLATION, AND LIPIDATION

Nitrosylation is the process in which proteins are taken up by a group of nitrous oxide particles, which affects protein activity and redox signaling [21]. ADP-ribosylation adds ADP-ribose units to all proteins, affecting processes such as DNA repair and

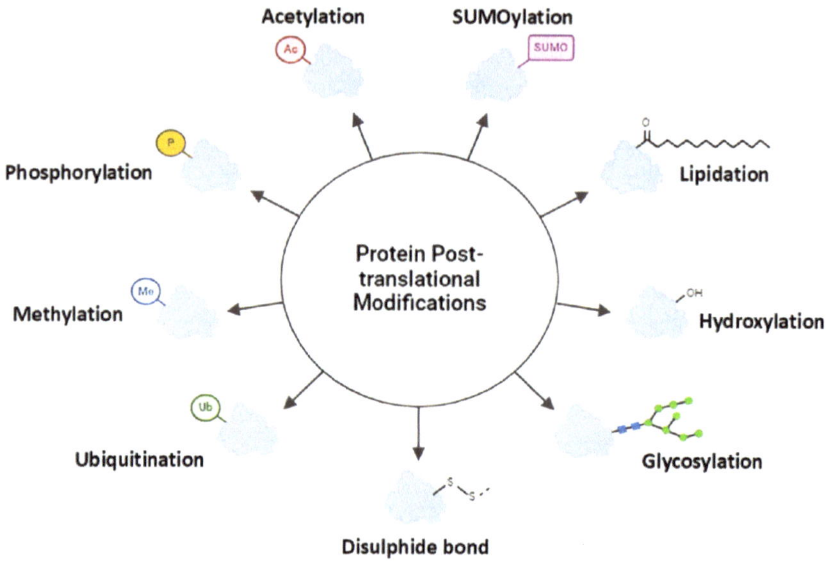

FIGURE 1.5 PTMs overview. (Created with `BioRender.com`.)

stress risk. The proteins anchor the membrane at the same time as the lipids, which modify the palm and oil lipids and do not control their position and function [36] (Figure 1.5). Each of these PTMs is essential for expanding the diverse functional repertoire of proteins. Understanding their mechanisms and roles is important to better understand complex cellular processes, disease mechanisms, and the possibility of therapeutic interventions.

1.3 ROLE OF PTMs IN DRUG DISCOVERY

PTMs and various diseases, including cancer, neurodegenerative diseases, and metabolic problems, have complex relationships. When PTMs are deregulated, they can cause inadequate cell signaling, changes in protein balance, and changes in gene expression, which can contribute to the development of various diseases [22]. Understanding PTMs is crucial in the field of drug development. It enables the development of new therapeutic approaches that focus on specific PTMs and the enzymes involved in the activity of PTMs. Drugs that can modulate PTMs are currently being investigated as potential treatments for diseases such as cancer, neuroprotection, and diseases related to the immune system [23].

1.4 MACHINE LEARNING AND COMPUTATIONAL MODELS FOR PTM PREDICTION

1.4.1 DEVELOPMENT OF A PTM PREDICTION MODEL

Figure 1.6 illustrates the outline steps of developing a model to predict PTMs. The first step is to specify the target PTM for prediction. Next is the data collection and preprocessing of the extracted sequences and relevant features. The third step is to choose a machine learning (ML) algorithm to develop and train the model using preprocessed data. The fourth step is evaluation of the model by assessing its performance and optimization of hyperparameters for improved performance. The fifth and final step is deployment of the model for prediction and continual updating with new experimental data for improvement.

1.4.2 APPLICATIONS OF THE CURRENT AND FORTHCOMING NEED FOR PTMs PREDICTION

PTM research is currently experiencing rapid growth, mainly due to advances in mass spectrometry, proteomics, and bioinformatics. However, experimental identification of PTM sites remains a costly and time-consuming task. This highlights the urgent need for computational techniques and prediction tools to accelerate the identification of PTM locations [24]. Hence, predictive models from computational

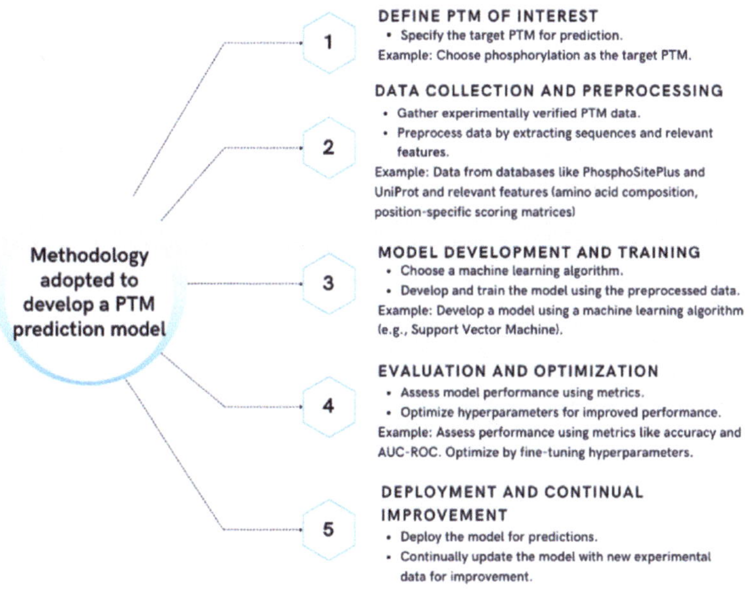

FIGURE 1.6 An outline of the process of developing a model to predict post-translational modifications.

biology and machine learning are increasingly popular in this environment. These models estimate the likely locations of PTMs within proteins from structural and sequence data, helping researchers focus their efforts on experimental validation [25]. Tables 1.1 and 1.2 list the databases and tools available for PTM prediction, respectively.

1.5 APPLICATIONS OF PTMs PREDICTION

1. **Drug Discovery and Development**: The use of PTM prediction in the realm of identifying and creating medications holds significant importance and is instrumental in several critical aspects. Firstly, it enables the identification and validation of potential drug targets by comprehensively understanding the involvement of PTMs in disease processes [24]. For instance, in diseases like cancer and Alzheimer's, PTM prediction has proven invaluable in pinpointing novel targets for therapeutic intervention. Additionally, PTM prediction facilitates the design of highly targeted therapies tailored to specific PTM profiles, thus ensuring a more precise and effective approach to treatment. Additionally, PTM prediction displays a critical role while optimizing efficacy of the drugs through strategies which are PTM guided, enhancing the overall impact of the therapeutic. PTM prediction also aids in monitoring progression of the disease by tracking the dynamic PTM profile of a patient's tumor over a period of time. This information then serves as a critically helpful tool for the assessment of the treatment response and detecting any indications of the progression of the disease [26–30].
2. **Personalized Medicine and Disease Diagnostics**: Over the years, PTM prediction has become pivotal to the advancement of personalized medicine and disease diagnostics. With comprehensive analysis of the PTM profile of a patient, healthcare providers can tailor treatment plans to optimize effectiveness and efficacy. This approach of predicting PTM of the proteins involved in the progression of a disease is particularly crucial in oncology. Here, understanding the PTM outline of a tumor microenvironment aids in selecting the most promising therapeutic agents. Moreover, it enables the identification of individuals at higher risk of developing specific diseases, by analysis of aberrant protein behavior. Thereby, allowing for closer monitoring and early intervention. For biomarker development, PTMs serve as invaluable indicators for disease diagnosis, prognosis, and response to treatment [31]. For instance, specific PTMs in tumor cells and their cellular microenvironment can serve as diagnostic markers for cancer, while the levels of certain PTMs can offer insights into a patient's likely response to chemotherapy, which targets those particular interacting proteins. The integration of PTM prediction further aids in the development of highly sensitive and specific biomarkers, potentially surpassing the accuracy, binding, and precision of current diagnostic tools [32].

TABLE 1.1
List of Databases for PTMs Prediction

Name	Type	Comments	Links
BioGRID	General	71 organisms and experiment primary	https://orcs.thebiogrid.org
CPLM	Lysine	122 organisms and additional experimental annotations	http://cplm.biocuckoo.org
dbOGAP	O-GlcNAcylated	–	https://bmcbioinformatics.biomedcentral.com/articles/10.1186/1471-2105-12-91
dbPSP	Phosphorylation	200 prokaryotic organisms and additional experimental annotations	http://dbpsp.biocuckoo.org
dbPTM	PTMs	–	https://awi.cuhk.edu.cn/dbPTM/
dbSNO	S-nitrosylation	18 organisms and additional experimental annotations	https://ngdc.cncb.ac.cn/databasecommons/database/id/70
DEPOD	Phosphorylation	Dephosphorylation interactions	https://www.depod.org
Eukaryotic Phosphorylation Site Database (EPSD)	Phosphorylation	Around 209,300 phosphoproteins from 68 eukaryotic organisms (24 plants, 7 protists, 18 animals, and 19 fungi) and additional experimental annotations data	http://epsd.biocuckoo.cn

(Continued)

TABLE 1.1 (*Continued*)
List of Databases for PTMs Prediction

Name	Type	Comments	Links
Expasy Proteomics Tools	Links of database and prediction servers	–	http://expasy.org/tools/#ptm
GlycoFish	Glycosylation	Zebrafish-related experimental and prediction data	http://betenbaugh.jhu.edu/GlycoFish
GlycoFly	Glycosylation	*Drosophila melanogaster* related experimental and prediction data	http://betenbaugh.jhu.edu/GlycoFly
HPRD	General	Human proteome initial additional experimental annotations	https://www.uniprot.org/uniprotkb?query=HPRD
hUbiquitome	Ubiquitylation	The premier database for human ubiquitination pathway proteins and their corresponding ubiquitin binding-site sequences.	https://pubmed.ncbi.nlm.nih.gov/22134927/
LymPHOS	Phosphorylation	Human and mouse experimental and predicted dataset	https://ngdc.cncb.ac.cn/databasecommons/database/id/4374

mUbiSiDa	Ubiquitylation	7 model organisms and additional experimental annotations	https://ngdc.cncb.ac.cn/databasecommons/database/id/1015
NetworKIN	Phosphorylation	Cellular phosphorylation networks	http://networkin.info/index.shtml
O-GlycBase	Glycosylation	Database for cross-species comparison	http://www.cbs.dtu.dk/databases/OGLYCBASE/
P3DB	Phosphorylation	45 plant organisms experimental and predicted data	http://www.p3db.org
PDBsite	General	Multiple locations within the three-dimensional structures of proteins	https://wwwmgs.bionet.nsc.ru/mgs/gnw/pdbsite/
PHOSIDA	General	9 organisms and additional experimental annotations	https://ngdc.cncb.ac.cn/databasecommons/database/id/610
PhosPhAT	Phosphorylation	*Arabidopsis thaliana*	http://phosphat.mpimp-golm.mpg.de
Phospho.ELM	Phosphorylation	Mainly model organisms and additional experimental annotations	http://phospho.elm.eu.org
Phospho3D	Phosphorylation	Mainly model organisms and additional experimental annotations	http://www.phospho3d.org
PhosphoBase	Phosphorylation	Database for validated information	http://phospho.elm.eu.org/

(*Continued*)

TABLE 1.1 (*Continued*)
List of Databases for PTMs Prediction

Name	Type	Comments	Links
PhosphoELM	Phosphorylation	Serine/Threonine/Tyrosine positions	http://phospho.elm.eu.org/
PhosphoNET	Phosphorylation	Human experimental and predicted data	http://www.phosphonet.ca
PhosphOrtholog	Phosphorylation	Database for cross-species comparison	http://phosphortholog.erc.monash.edu/
Phosphosite Plus	General	26 organisms and initial experimental data	https://www.phosphosite.org
PLMD	General	176 organisms and additional experimental annotations	http://plmd.biocuckoo.org
PROSITE	General	Consensus patterns–based curation	http://www.expasy.org/prosite/
pTestis	Phosphorylation	Mouse and experimental and predicted	http://ptestis.biocuckoo.org
PTMCode v2	General	19 organisms with additional experimental annotations	http://ptmcode.embl.de
PTM-SD	General	7 model organisms with additional experimental annotations	https://ngdc.cncb.ac.cn/databasecommons/database/id/862

PupDB	Pupylation	268 proteins identified as pupylated, featuring 311 confirmed pupylation sites, alongside 1123 proteins that are potential candidates for pupylation	https://link.springer.com/content/pdf/10.1186/1471-2105-13-40.pdf
qPTM	General	Human additional experimental annotations	http://qptm.omicsbio.info/
RegPhos	Phosphorylation	Human, mouse, and rat experimental and predicted data	https://catalog.integbio.jp/dbcatalog/en/record/nbdc01681
RESID	General	–	https://proteininformationresource.org/resid/
SwissPalm	S-palmitoylation	17 organisms with experimental and prediction data	https://swisspalm.org
SysPTM	General	6 organisms with additional experimental annotations	https://ngdc.cncb.ac.cn/databasecommons/database/id/4538
UbiNet	Ubiquitylation	E3 binding/recognition site	https://openebench.bsc.es/tool/ubine
UniPep	Glycosylation	Human experimental and predicted data	https://ngdc.cncb.ac.cn/databasecommons/database/id/3146
UniProt	General	Highly curated and annotated	https://www.uniprot.org

(Continued)

TABLE 1.1 (*Continued*)
List of Databases for PTMs Prediction

Name	Type	Comments	Links
WERAM	Histone acetylation, and methylation	148 eukaryotic species	http://weram.biocuckoo.org/index.php
YAAM	General	*Saccharomyces cerevisiae* and additional experimental annotations	http://yaam.ifc.unam.mx

TABLE 1.2
List of Tools For PTM Prediction

Name	Type	Extra Information	Links
AMS 4.0	General	Consensus prediction	http://code.google.com/p/automotifserver
ASEB (Acetylation Set Enrichment Based)	KAT-specific acetylation	Data derived from protein-protein interactions	https://biomedinfo.bjmu.edu.cn/swxxzy/7e1ab51d045247ef83de23594e897b93.htm
big-PI/GPI animals	Lipid	Glycosylphosphatidylinositol (GPI) modification sites identified in animal species	http://mendel.imp.ac.at/gpi/gpi_server.html
big-PI/GPI fungi	Lipid	GPI modification sites identified in fungal species	http://mendel.imp.ac.at/gpi/fungi_server
big-PI/GPI plants	Lipid	GPI modification sites identified in plant species	https://mendel.imp.ac.at/gpi/plant_server
CKSAAPOGlySite	Glycosylation	Mammalian only	https://mybiosoftware.com/cksaap_oglysite-predictor-of-mucin-type-o-glycosylation-sites.html
CKSAAPUbSite	Ubiquitination	Support Vector machine (SVM) based	http://systbio.cau.edu.cn/cksaap_ubsite/index.php?page=introduction

(Continued)

TABLE 1.2 (*Continued*)
List of Tools For PTM Prediction

Name	Type	Extra Information	Links
CKSAAP-Palm	Palmitoylation	SVM based	http://doc.aporc.org/wiki/CKSAAP-Palm
DeepAcet	Lysine acetylation	Neural Network based	https://github.com/Sunmile/DeepAcet
DeepIPS	Phosphorylation	Ensemble of Long short-term memory (LSTM) and convolutional neural networks (CNNs)for SARS-CoV-2	http://lin-group.cn/server/DeepIPs/
DeepPhos	Phosphorylation	CNN	https://github.com/USTC-HIlab/DeepPhos
DictyOGlyc	Glycosylation	Sites of O-(alpha)-GlcNAc glycosylation identified in proteins, trained specifically on *Dictyostelium discoideum*	https://services.healthtech.dtu.dk/service.php?DictyOGlyc-1.1

DNNAce	Acetylation	Identification of lysine acetylation sites in prokaryotes using deep neural networks that integrate multiple sources of information	`https://github.com/QUST-AIBBDRC/DNNAce/`
ESA-UbiSite	Ubiquitination	ESA-based prediction method	`https://academic.oup.com/bioinformatics/article/33/5/661/2849433`
FindMod	General	Along with normal prediction, single amino acid substitutions are also predicted	`https://web.expasy.org/findmod/`
GlycoMod	N-,O-glycosylations	Predicts probable oligosaccharide structures on protein by using their experimental data	`https://web.expasy.org/glycomod/`
GLYCOPP	N-,O-glycosylations	SVM with different methods such as CPP, PPP, BPP, ASA+BPP	`http://crdd.osdd.net/raghava/glycopp/submit.html`

(*Continued*)

TABLE 1.2 (*Continued*)
List of Tools For PTM Prediction

Name	Type	Extra Information	Links
GPP	N-,C-,O-glycosylations	Both sites and tendency of association with modified residues by Random forest	`https://comp.chem.nottingham.ac.uk/glyco/`
GPS 6.0	Phosphorylation	Kinase-specific phosphorylation sites	`https://gps.biocuckoo.cn/`
GPS-SNO	S-nitrosylation	Based on 504 experimentally verified SNOs	`http://sno.biocuckoo.org/online.php`
GPS-SUMO	SUMOylation	Novel SIMs prediction feature using penalized logistic regression	`http://sumosp.biocuckoo.org/`
HydPred	Hydroxylation	The combination of SMOTE, a method for oversampling minority data, paired with the Random Forest algorithm	`http://cppdd.cn/hydpred/`

IMP-PUP	Pupylation	An adapted self-training SVM method tailored for prokaryotic data with a semi-supervised approach	`https://juzhe1120.github.io/`
iPhoPred	Phosphorylation	SVM based	`http://lin-group.cn/server/iPhoPred/`
iPUP	Pupylation	Sequence-based prediction	`https://pubmed.ncbi.nlm.nih.gov/23871866/`
JASSA (Joined Advanced SUMOylation Site and Sim Analyzer)	SUMOylation	Analysis of Advanced SUMOylation site and SIMs by utilizing positional frequency matrix analysis containing experimental data	`http://www.jassa.fr`
KinasePhos	Phosphorylation	Kinase-specific phosphorylation	`https://github.com/tom-209/KinasePhos-3.0-executable-file`
MPSite	Phosphorylation	Microbial phosphorylation	MPSite (kyutech.ac.jp)
MusiteDeep	General	Prediction and visualization	`https://www.musite.net/`

(Continued)

TABLE 1.2 (*Continued*)
List of Tools For PTM Prediction

Name	Type	Extra Information	Links
Myristoylator	N-myristoylated	Neural network, an additional tool for NMT	`http://web.expasy.org/myristoylator/`
MyrPS/NMT	Lipid	Myristoyl-CoA:protein N-myristoyltransferase	`https://mendel.imp.ac.at/myristate/SUPLpredictor.htm`
NetAcet	Nα-acetylated	Yeast and mammalian site prediction using Neural network	`https://services.healthtech.dtu.dk/service.php?NetAcet-1.0`
NetCGlyc	Glycosylation	C-mannosylation locations in mammalian proteins	`https://services.healthtech.dtu.dk/service.php?NetCGlyc-1.0`
NetGlycate	Glycosylation	The process of glycation affecting the ε amino groups of lysines within proteins found in mammals	`https://services.healthtech.dtu.dk/service.php?NetGlycate-1.0`
NetNGlyc	Glycosylation	N-linked glycosylation sites in human proteins built on neural networks	`https://services.healthtech.dtu.dk/service.php?NetNGlyc-1.0`

NetOGlyc	Glycosylation	Glycosylation sites characterized by O-GalNAc (mucin type) modification observed in eukaryotic mammalian proteins	https://services.healthtech.dtu.dk/service.php?NetOGlyc-4.0
NetPhos	Phosphorylation	Eukaryotic proteins data	https://services.healthtech.dtu.dk/service.php?NetPhos-3.1
NetPhosBac	Phosphorylation	Bacterial	https://services.healthtech.dtu.dk/service.php?NetPhosBac-1.0
NetPhospan	Phosphorylation	CNNs	https://services.healthtech.dtu.dk/service.php?NetPhospan-1.0
NetPhosYeast	Phosphorylation	Sites of phosphorylation specifically on serine and threonine residues within proteins found in yeast	https://services.healthtech.dtu.dk/service.php?NetPhosYeast-1.0
PeroPS/ PTS1	The peroxisomal targeting signal 1	–	https://mendel.imp.ac.at/pts1/
Phobius	Signal peptide and transmembrane topology predictor	–	http://phobius.sbc.su.se

(*Continued*)

TABLE 1.2 (*Continued*)
List of Tools For PTM Prediction

Name	Type	Extra Information	Links
PhosIDN	Phosphorylation	Integrating both sequence data and protein-protein interaction details for prediction purposes	`https://github.com/ustchangyuanyang/PhosIDN`
PhosphoSVM	Phosphorylation	SVM based at S/T/Y amino acids	`http://sysbio.unl.edu/PhosphoSVM/prediction.php`
PhosPred-RF	Phosphorylation	Sequence based	`https://pubmed.ncbi.nlm.nih.gov/28166503/`
PhosTransfer	Phosphorylation	Kinase-specific phosphorylation sites predicted using deep transfer learning with CNNs hierarchally	`https://github.com/yxu132/PhosTransfer`
PrePS	Lipid	Prenylation	`https://mendel.imp.ac.at/PrePS/`
PSuccE	Succinylation	Ensemble SVM	`https://github.com/ningq669/PSuccE`
PTMPred	General	PSPMs used to calculate Kernel matrix with SVM	`http://doc.aporc.org/wiki/PTMPred`
PTMscape	General	Phosphorylation and ubiquitination along with map modifications	`https://github.com/ginnyintifa/PTMscape`

PUL-PUP	Pupylation	Positive-unlabeled learning technique	https://pul-pup.github.io/
PupS	Pupylation	In prokaryotic proteins, extreme learning machine	http://www.csbio.sjtu.edu.cn/bioinf/PupS/
Scansite	Phosphorylation	Motifs present within proteins that are predicted or likely to undergo phosphorylation by specific protein kinases	https://scansite4.mit.edu
SeqPalm	Palmitoylation	S-palmitoylation, synthetic minority oversampling technique	http://cppdd.cn/seqpalm/
SignalP	Cleavage sites and signal peptide	Gram-positive, Gram-negative, and eukaryotic amino acid sequences	https://services.healthtech.dtu.dk/service.php?SignalP-5.0
SuccinSite	Succinylation	Random forest classifier	https://pubmed.ncbi.nlm.nih.gov/26739209/
SulCysSite	S-sulfenylation	Random forest classifier	http://kurata14.bio.kyutech.ac.jp/SulCysSite/
SVM-SulfoSite	Sulfenylation	SVM	https://github.com/HussamAlbarakati/SVM-Sulfosite

(Continued)

TABLE 1.2 (*Continued*)
List of Tools For PTM Prediction

Name	Type	Extra Information	Links
TargetP	Presence of N-terminal presequences	Mitochondrial, chloroplastic, secretory pathway, or other	`https://services.healthtech.dtu.dk/service.php?TargetP-2.0`
UbiSitePred	Ubiquitination	Lasso	`https://github.com/QUST-AIBBDRC/UbiSitePred/`
UbPred	Ubiquitination	Random forest	`http://www.ubpred.org`
YinOYang	N-,C-,O-glycosylations	Using Yin (cellular)–Yang (nuclear) proteins	`https://services.healthtech.dtu.dk/service.php?YinOYang-1.2`

Abbreviations: ASA, Average Surface Accessibility; BPP, Binary profile of pattern; CNN, convolutional neural network; CPP, Composition profile of patterns; GPI, glycosylphosphatidylinositol; KAT, (K)-acetyl transferase; LSTM, long short-term memory; NMT, neural machine translation; PPP, PSSM profile of pattern; PSPMs, position-specific propensity matrices; SARS-CoV-2, severe acute respiratory syndrome coronavirus 2; SIMs, SUMO-interacting motifs; SMOTE, synthetic minority oversampling technique; SNOs, S-nitrosothiols; S/T/Y, Serine/Threonine/Tyrosine; SVM, support vector machine.

3. **Targeted Functional Genomics**: PTM prediction is a powerful tool in the field of targeted functional genomics research and development. It provides a means to explore the impact and significance of PTMs and plays an indispensable role in fundamental biological processes like cell signaling, gene regulation, and protein transport. This understanding of how PTMs in a complex manner regulate protein function creates new avenues into the investigation of the molecular intricacies of how proteins regulate life through more than just their structure, but also by external modifications. Through the prediction of PTM sites, researchers can unveil the functional implications and involvement within specific pathways or disease pathology, shedding light on the composite interplay between PTMs and cellular activities. Thus, this application of PTM prediction revolutionizes our comprehension of how post-translational modifications shape biological systems at a molecular level and how functional genomics play a part in it [33].

1.6 CHALLENGES FACED DURING DEVELOPMENT OF PREDICTION MODELS

1. **Data quality and availability**: The in vitro physical experimental techniques for discovering and identification of PTMs are resource and time inefficient, resulting in limited and noisy data for training computational methods, thus taking years, labor, and funds to study just a handful of proteins. It is imperative that development of standardized testing datasets that can provide a fair and consistent benchmark for comparing different prediction tools is a need of the hour [34]. There is also a lack of standardization and annotation of PTM data across different databases and resources, which hinders not only production of quality datasets but brings inconsistency in molecular subtext, functionality, and cellular activity they take part in [35].
2. *Sequence diversity and complexity*: PTMs may take place on various amino acids, thereby presenting diverse features and modifications, occurring in multiple combinations and contexts. The PTMs are influenced by numerous factors, such as protein structure, interactions, environment, and evolution, which are not captured by local sequence fragments alone, thereby needing more context during model development [36]. The sequence patterns and features associated with PTMs lack a fixed or decipherable scheme and are diverse and complex, making it difficult to capture them by simple rules or motifs. The integration of protein sequence and structural data poses computational and methodological difficulties in a system, especially for modeling 3D spatial information and long-range interactions [37]. Data filtering, which is essential for avoiding overfitting, lacks consensus on criteria due to variable data availability, while dataset balancing, which is crucial to mitigate biases in imbalanced positive and negative datasets, is also

absent [5]. Hence, machine learning models need to incorporate additional information and features to improve the accuracy and interpretability of PTM prediction.

3. **Classification performance and generalization**: The computational methods for predicting PTMs usually exhibit low accuracy, precision, and recall, especially for rare or novel PTMs. The algorithms also have a tendency to overfit the training data and consequently fail to generalize to new or unseen sequences or PTMs [5]. Feature extraction, a vital step in the process of predictor learning, is hampered while selecting informative biological descriptors for encoding protein sequences into numerical feature vectors, due to poor data quality and lack of descriptors, among many [5].
4. **Validation and interpretation**: The prediction model of PTMs needs to be validated by experimental methods, which are costly in both time and resource, for instance, in vitro testing. The predictions also need to be interpreted in terms of their biological functions and implications, which are not always clear or well-understood, thanks to lack of information or in the case of novel PTMs [38]. The interpretation and validation of prediction results are required to be effective and biologically meaningful, along with the ability to explain the model's decision and verification of the predicted PTMs with experimental data [37].

1.7 FUTURE PERSPECTIVES IN THE DOMAIN OF PTMs PREDICTION

Some possible future perspectives that PTM prediction has are:

1. **Improvement of the accuracy and efficiency of PTM prediction methods:** Current PTM prediction methods still face some challenges, including an imbalance between positive and negative samples, the diverse and intricate nature of PTM types, the absence of reliable and comprehensive PTM databases, and the integration of multiple sources of information. Developing more advanced machine learning and deep learning techniques, such as ensemble learning, transfer learning, attention mechanism, and graph neural networks, may help to overcome these challenges and improve the performance of PTM prediction methods [39].
2. **Expansion of the coverage and scope of PTM prediction methods:** Most of the existing PTM prediction methods focus on a few common PTM types, such as phosphorylation, ubiquitination, and glycosylation. However, there are many other PTM types that have not been well studied, such as methylation, acetylation, and SUMOylation. Developing more PTM prediction methods for these less-studied PTM types may help to reveal their biological functions and roles in diseases [40]. Moreover, most of the PTM prediction methods are designed for specific species, such as human, mouse, or yeast. Developing more cross-species PTM prediction methods may help to explore the evolutionary and comparative aspects of PTMs.

3. **PTM prediction methods:** The deep examination of protein interactions and regulations using PTM prediction models gives detailed insight on protein functions and cellular processes. The development of more PTM prediction models and algorithms may assist to explore the complexity of PTM networks and their regulations. There are various factors involved in the regulation of PTMs like enzymes, cofactors, substrates, and various environmental conditions. So it is advisable to incorporate all these factors in developing PTM prediction models. This will help in elucidating the mechanisms and dynamics of PTMs [5].
4. **Functional annotation and network analysis:** Functional annotation and network analysis depend on protein primary sequence and crystallographic structure. Hence, future research should incorporate PTM prediction models into functional annotation and network analysis to help in deciphering the functional and regulatory roles of PTMs across diverse biological conditions.
5. Additionally, the integration of PTM prediction with other bioinformatics tools and applications emerges as a trend to help in addressing diverse biological and biomedical challenges, including protein function annotation, prediction of protein-protein interactions, and drug discovery. Future endeavors in this field can include exploring the synergistic integration and inclusion of PTM prediction with existing bioinformatics tools and workflows, thus, aiming to offer more comprehensive and precise insights and valuable information into the molecular mechanisms and functions of proteins [36].
6. The biggest challenge yet is the acknowledgment of a gap between the prediction and interpretation of PTMs and addressing the need for future research to extend beyond technical considerations and dive into the biological and evolutionary significance of PTMs in organisms. The forthcoming studies should aim to study the impact of PTMs on protein structure, dynamics, interactions, and evolution, while also examining the regulatory influences exerted by various cellular and environmental factors on them [34].

1.8 CASE STUDY: EVALUATION OF PTM PREDICTION TOOLS USING BENCHMARK DATASETS

The use of machine learning methodologies to predict PTM sites showcases an effective strategy to simplify the identification process. However, despite the extensive content available on PTM-related databases like dbPTM and PhosphoSitePlus [41], which comprise information from research articles and public resources, precise and accurate prediction methods are needed to effectively filter and screen potential PTM sites. This case study evaluates the performance of specific prediction tools concerning various PTM types using benchmark datasets obtained from dbPTM [42].

1.8.1 METHODS

1. Dataset Selection
   - Accessing dbPTM, benchmark datasets for distinct PTM types were downloaded, including acetylation, phosphorylation by cyclin-dependent kinases (CDK), S-palmitoylation (for lipid modification), SUMOylation, and N-linked glycosylation.
   - Acetylation, phosphorylation by CDK, and N-linked glycosylation datasets comprised both positive and negative samples, whereas the remaining datasets contained only positive samples.
2. Dataset Preparation: The first 20 human FASTA sequences were selected from each dataset.
3. Tool Selection: Specific prediction tools were chosen for analysis
   - GPS-PAIL for acetylation
   - GPS 6.0 for phosphorylation
   - GPS-Lipid for S-palmitoylation
   - JASSA v4 for SUMOylation
   - NetNGlyc-1.0 for N-glycosylation
4. Tool Execution and Prediction Process
   - Each dataset was subjected to its respective prediction tool.
   - For phosphorylation, only CDK-specific phosphorylation sites were selected for prediction; likewise, only palmitoylation sites were chosen for lipid modification prediction and lysine acetylation for acetylation prediction.
   - Results were recorded based on the ability of each tool to predict the designated PTM types. Each dataset was run on its respective tool, focusing on specific modifications (e.g., CDK phosphorylation, lysine acetylation).
5. Performance Metrics Calculation and Evaluation
   - Accuracy, precision, and specificity were calculated for datasets containing both positive and negative samples (acetylation, phosphorylation by CDK, and N-linked glycosylation).
   - Recall (sensitivity) was computed for all datasets.
   - Accuracy denotes how correctly a model predicts true positive values, that is, determines if a PTM is really taking place in a protein. It was calculated by the formula:

$$Accuracy = \frac{Number\ of\ correct\ predictions}{Total\ number\ of\ samples} \tag{1.1}$$

The meaning of terms used below are as follows:

True Positives: Number of proteins predicted to have PTMs, but actually having PTMs.

True Negatives: Number of proteins predicted to not have PTMs, but actually lacking PTMs.

False Positives: Number of proteins predicted to have PTMs, but actually lacking PTMs.

False Negatives: Number of proteins predicted to not have PTMs, but actually having PTMs.

Precision denotes how closely the positive predictions, that is, the presence of PTMs made by the model, are in reality are present in the protein. It was calculated by the formula:

$$Precision = \frac{True\ Positives}{True\ Positives + False\ positives} \tag{1.2}$$

Specificity describes how correctly a model is able to predict real or true negatives, that is, the absence of PTMs in the proteins, when compared to all the unmodified proteins in the dataset. It was calculated by the formula:

$$Specificity = \frac{True\ Negatives}{True\ Negatives + False\ Positives} \tag{1.3}$$

Sensitivity describes how closely a model is able to predict real or true positive values, that is, the presence of PTMs in proteins, when compared to all the proteins that have the respectively tested modifications. It was calculated by the formula:

$$Sensitivity = \frac{True\ Positives}{True\ Positives + False\ Negatives} \tag{1.4}$$

1.8.2 RESULTS

Table 1.3 reveals calculated prediction scores of various PTM types. The accuracy, precision, and specificity scores could not be calculated for lipid and SUMOylation due to absence of negative datasets. Best performance was indicated for prediction of N-glycosylation by NetNGlyc-1.0 with highest scores for three out of four performance parameters.

TABLE 1.3

The Accuracy, Precision, Specificity and Recall (Sensitivity) Scores for Prediction of Various PTM Types

PTM Type	Accuracy	Precision	Specificity	Recall (Sensitivity)
Lipid	–	–	–	0.75
Acetylation	0.75	0.692	0.6	0.9
N-glycosylation	0.875	0.941	0.95	0.8
SUMOylation	–	–	–	0.75
Phosphorylation	0.825	0.81	0.8	0.85

1.8.3 INTERPRETATION

- **Acetylation:** The tool displayed moderate accuracy and specificity but exhibited high sensitivity and acceptable precision, indicating better prediction of acetylation sites.
- **Phosphorylation:** Demonstrated good overall performance with high accuracy, specificity, precision, and sensitivity in predicting CDK phosphorylation sites.
- **S-Palmitoylation aad SUMOylation:** Limited results but showed moderate sensitivity.
- **N-Glycosylation:** Showed high accuracy, specificity, precision, but lower sensitivity in prediction of N-glycosylation sites.

1.9 CONCLUSION

In conclusion, PTM prediction stands as a challenging yet proficient arena in discovering the underlying complexities, significance, and interactions of molecular modifications. The pursuit of development of robust predictive methodologies for PTMs, coupled with advancements in bioinformatics tools and databases, display a bright future paving the way for understanding dynamics and mechanisms of the biological world, life. In the near future, there is a requirement for progress in existing tools for PTM prediction. The prospects lie in improving predictive methodologies and algorithms by leveraging advanced machine learning techniques, while also broadening the scope of study for prediction to include less-studied PTM types, and finally, unveiling to understand the interplay and regulations among different PTM types. The case study evaluating PTM prediction tools with the help of benchmark datasets demonstrated the varying performance metrics of the prediction tools evaluated across varying PTM types. The study revealed high sensitivity but variable accuracy and precision in the different models. The case study sheds light on the potential gaps and improvement prospects within existing prediction tools, highlighting the necessity for continued research for more comprehensive algorithms, building more accurate and precise models, their refinement, and further enhancement.

REFERENCES

1. Christoph Kannicht and Beate Fuchs. Post-translational modifications of proteins. In *Molecular Biomethods Handbook: Second Edition*. Editors: John M. Walker (Professor Emeritus), Ralph Rapley, 427–449. Springer Protocols Handbooks. Humana Press, 2008.
2. Vladimir N Uversky. Posttranslational modification. In *Brenner's Encyclopedia of Genetics: Second Edition*, Editors Stanley Maloy and Kelly Hughes, pp. 425–430. 2013.
3. Monika Khandelwal and Ranjeet Kumar Rout. Prmxai: protein arginine methylation sites prediction based on amino acid spatial distribution using explainable artificial intelligence. *BMC Bioinformatics*, 24(1):376, 2023.

4. Boris Nikolaevich Sobolev, Aleksandr Vladimirovich Veselovsky, and Vladimir Vasil'evich Poroikov. Prediction of protein post-translational modifications: main trends and methods. *Russian Chemical Reviews*, 83(2):143, 2014.
5. Shahin Ramazi and Javad Zahiri. Post-translational modifications in proteins: resources, tools and prediction methods. *Database*, 2021:baab012, 2021.
6. Francois-Xavier Theillet, Caroline Smet-Nocca, Stamatios Liokatis, Rossukon Thongwichian, Jonas Kosten, Mi-Kyung Yoon, Richard W Kriwacki, Isabelle Landrieu, Guy Lippens, and Philipp Selenko. Cell signaling, post-translational protein modifications and nmr spectroscopy. *Journal of biomolecular NMR*, 54:217–236, 2012.
7. Chun-Wei Tung and Shinn-Ying Ho. Computational identification of ubiquitylation sites from protein sequences. *BMC Bioinformatics*, 9:1–15, 2008.
8. Anand Patwardhan, Norton Cheng, and JoAnn Trejo. Post-translational modifications of g protein–coupled receptors control cellular signaling dynamics in space and time. *Pharmacological Reviews*, 73(1):120–151, 2021.
9. Inmaculada Navarro-Lérida, Miguel Sánchez-Álvarez, and Miguel Ángel Del Pozo. Post-translational modification and subcellular compartmentalization: emerging concepts on the regulation and physiopathological relevance of rhogtpases. *Cells*, 10(8):1990, 2021.
10. Helena Ryšlavá, Veronika Doubnerová, Daniel Kavan, and Ondřej Vaněk. Effect of post-translational modifications on enzyme function and assembly. *Journal of Proteomics*, 92:80–109, 2013.
11. Joerg Reinders and Albert Sickmann. Modificomics: posttranslational modifications beyond protein phosphorylation and glycosylation. *Biomolecular Engineering*, 24(2):169–177, 2007.
12. Wolfgang Eberhardt, Anke Doller, and Josef Pfeilschifter. Regulation of the mRNA-binding protein hur by posttranslational modification: spotlight on phosphorylation. *Current Protein and Peptide Science*, 13(4):380–390, 2012.
13. Nikolaj Blom, Thomas Sicheritz-Pontén, Ramneek Gupta, Steen Gammeltoft, and Sren Brunak. Prediction of post-translational glycosylation and phosphorylation of proteins from the amino acid sequence. *Proteomics*, 4(6):1633–1649, 2004.
14. Md Mehedi Hasan and Mst Shamima Khatun. Prediction of protein post-translational modification sites: an overview. *Annals of Proteomics and Bioinformatics*, 2(1):49–57, 2018.
15. Monika Khandelwal, Ranjeet Kumar Rout, Saiyed Umer, Saurav Mallik, and Aimin Li. Multifactorial feature extraction and site prognosis model for protein methylation data. *Briefings in Functional Genomics*, 22(1):20–30, 2023.
16. Monika Khandelwal and Ranjeet Kumar Rout. Deepprms: advanced deep learning model to predict protein arginine methylation sites. *Briefings in Functional Genomics*, page elae001, 2024.
17. David G Christensen, Xueshu Xie, Nathan Basisty, James Byrnes, Sean McSweeney, Birgit Schilling, and Alan J Wolfe. Post-translational protein acetylation: an elegant mechanism for bacteria to dynamically regulate metabolic functions. *Frontiers in Microbiology*, 10:469204, 2019.
18. Verónica Alonso and Peter A Friedman. Minireview: ubiquitination-regulated g protein-coupled receptor signaling and trafficking. *Molecular Endocrinology*, 27(4):558–572, 2013.
19. Annette Flotho and Frauke Melchior. Sumoylation: a regulatory protein modification in health and disease. *Annual Review of Biochemistry*, 82:357–385, 2013.

20. Natalie Foot, Tanya Henshall, and Sharad Kumar. Ubiquitination and the regulation of membrane proteins. *Physiological Reviews*, 97(1):253–281, 2017.
21. Robert SB Clark, Hulya Bayir, and Larry W Jenkins. Posttranslational protein modifications. *Critical Care Medicine*, 33(12):S407–S409, 2005.
22. Fanwang Meng, Zhongjie Liang, Kehao Zhao, and Cheng Luo. Drug design targeting active posttranslational modification protein isoforms. *Medicinal Research Reviews*, 41(3):1701–1750, 2021.
23. Lin-hui Zhai, Kai-feng Chen, Bing-bing Hao, and Min-jia Tan. Proteomic characterization of post-translational modifications in drug discovery. *Acta Pharmacologica Sinica*, 43(12):3112–3129, 2022.
24. Alexandre G de Brevern and Joseph Rebehmed. Current status of ptms structural databases: applications, limitations and prospects. *Amino Acids*, 54(4):575–590, 2022.
25. A Saskia Venne, Laxmikanth Kollipara, and René P Zahedi. The next level of complexity: crosstalk of posttranslational modifications. *Proteomics*, 14(4-5):513–524, 2014.
26. Pooja Gulhane and Shailza Singh. Unraveling the post-translational modifications and therapeutical approach in nsclc pathogenesis. *Translational Oncology*, 33:101673, 2023.
27. Madhu Ramesh, Pushparathinam Gopinath, and Thimmaiah Govindaraju. Role of post-translational modifications in alzheimer's disease. *ChemBioChem*, 21(8):1052–1079, 2020.
28. Michael Schauperl and Rajiah Aldrin Denny. AI-based protein structure prediction in drug discovery: impacts and challenges. *Journal of Chemical Information and Modeling*, 62(13):3142–3156, 2022.
29. Neha Varshney and Abhinava K Mishra. Deep learning in phosphoproteomics: methods and application in cancer drug discovery. *Proteomes*, 11(2):16, 2023.
30. Huimin Zhang, Jixiao He, Guang Hu, Fei Zhu, Hao Jiang, Jing Gao, Hu Zhou, Hua Lin, Yingjuan Wang, Kaixian Chen, et al. Dynamics of post-translational modification inspires drug design in the kinase family. *Journal of Medicinal Chemistry*, 64(20):15111–15125, 2021.
31. Trevor T Duarte and Charles T Spencer. Personalized proteomics: the future of precision medicine. *Proteomes*, 4(4):29, 2016.
32. Dana Pascovici, Jemma X Wu, Matthew J McKay, Chitra Joseph, Zainab Noor, Karthik Kamath, Yunqi Wu, Shoba Ranganathan, Vivek Gupta, and Mehdi Mirzaei. Clinically relevant post-translational modification analyses—maturing workflows and bioinformatics tools. *International Journal of Molecular Sciences*, 20(1):16, 2018.
33. Yujuan Deng, Yunfang Fu, Huitao Zhang, Xin Liu, and Zhiguo Liu. Protein post-translational modification site prediction using deep learning. *Procedia Computer Science*, 198:480–485, 2022.
34. Brett Trost and Anthony Kusalik. Computational prediction of eukaryotic phosphorylation sites. *Bioinformatics*, 27(21):2927–2935, 2011.
35. Subhadip Basu and Dariusz Plewczynski. Ams 3.0: prediction of post-translational modifications. *BMC Bioinformatics*, 11:1–15, 2010.
36. Marcin Tatjewski, Marcin Kierczak, and Dariusz Plewczynski. Predicting post-translational modifications from local sequence fragments using machine learning algorithms: Overview and best practices. *Prediction of Protein Secondary Structure*, Editors: Yaoqi Zhou, Andrzej Kloczkowski, Eshel Faraggi, Yuedong Yang, pp. 275–300, Humana Press, 2017.
37. Yu Yan, Jyun-Yu Jiang, Mingzhou Fu, Ding Wang, Alexander R Pelletier, Dibakar Sigdel, Dominic CM Ng, Wei Wang, and Peipei Ping. Mind-s is a deep-learning

prediction model for elucidating protein post-translational modifications in human diseases. *Cell Reports Methods*, 3(3), 2023.

38. Damiano Piovesan, Andras Hatos, Giovanni Minervini, Federica Quaglia, Alexander Miguel Monzon, and Silvio CE Tosatto. Assessing predictors for new post translational modification sites: A case study on hydroxylation. *PLoS Computational Biology*, 16(6):e1007967, 2020.
39. Kuo-Chen Chou. Progresses in predicting post-translational modification. *International Journal of Peptide Research and Therapeutics*, 26(2):873–888, 2020.
40. Duolin Wang, Dongpeng Liu, Jiakang Yuchi, Fei He, Yuexu Jiang, Siteng Cai, Jingyi Li, and Dong Xu. Musitedeep: a deep-learning based webserver for protein post-translational modification site prediction and visualization. *Nucleic Acids Research*, 48(W1):W140–W146, 2020.
41. Cheng-Tsung Lu, Kai-Yao Huang, Min-Gang Su, Tzong-Yi Lee, Neil Arvin Bretana, Wen-Chi Chang, Yi-Ju Chen, Yu-Ju Chen, and Hsien-Da Huang. Dbptm 3.0: an informative resource for investigating substrate site specificity and functional association of protein post-translational modifications. *Nucleic Acids Research*, 41(D1):D295–D305, 2013.
42. Zhongyan Li, Shangfu Li, Mengqi Luo, Jhih-Hua Jhong, Wenshuo Li, Lantian Yao, Yuxuan Pang, Zhuo Wang, Rulan Wang, Renfei Ma, et al. dbptm in 2022: an updated database for exploring regulatory networks and functional associations of protein post-translational modifications. *Nucleic Acids Research*, 50(D1):D471–D479, 2022.

2 Application of Artificial Intelligence in Recognition of Gene Regulation and Metabolic Pathways

Mayank Roy Chowdhury
and Sudarshana Deepa Vijaykumar

2.1 INTRODUCTION

The orchestration of gene expression and metabolic pathways is central to the "digital" logic of molecular biology, conferring upon cells the "programmability" to respond to changing environmental contingencies and extrinsic stimuli, as well as a means for sustaining basic life processes [1]. This chapter, the intricacies of genomic regulation and metabolic pathways are discussed. The fine-tuning of transcription factors—the regulatory switches of biological functions—requires thousands of genetic and biochemical interactions among competing regulatory proteins. Within a typical cell, the biochemical machinery is a coordinated series of enzymatic reactions spanning thousands of complex molecular steps. Beneath the surface of an integrated biological function, a teeming web of complexity is often more challenging to track than can be handled by traditional analytical approaches [2]. Several constraints make it difficult to investigate the dynamics of gene regulation using experimental techniques and obstruct the way to a thorough knowledge [3]. Still, even with the great potential of post-advanced microscopy, the complexity of controlled gene expression is often shrouded in mystery due to the nature of DNA, which has a diameter measuring merely 2 nanometers and a flexibility that can be compared to a live performer in a busy city. This brings us to the problem of what may be considered a scale at a few atoms wide It presents itself as a challenging task because the actual observation of molecules going through the complicated dance called gene regulation still eludes us [4, 5]. Another level of intricacy is added by the dynamic nature of DNA and its inclination to go through quick and complicated changes, including

DOI: 10.1201/9781032660714-2

looping, folding, and twisting [6, 7]. There is a collection of methods where talented researchers can find the relevant patterns in huge datasets and explain the complex relations that regulate these biological processes. Machine learning algorithms, including neural networks and computational techniques using artificial intelligence capability, have provided a new perspective on the networks that govern gene control and metabolic rates, thereby transforming the unknown into information [8–11].

Though the techniques of high-throughput sequencing have made immense contributions to obtaining substantial genomic data, giving light to complex gene regulatory mechanisms, they present problems in turn. There is an accumulation of massive datasets providing a peephole into this intricate interplay of genes; however, it is quite challenging to make any sense of the sea of information collected that requires sophisticated computational methods [12]. These analyses are thus limited by the very size and structure of the data and by the fact that noise and systematic biases may complicate the datasets [13]. This complicated situation is also a driving force for researchers who seek to explain the mysterious "Gene regulation" and explore the new frontiers of molecular biology with progressive strategies and technologies.

2.1.1 ARTIFICIAL INTELLIGENCE IN BIOLOGY

Integrating two disciplines has proven to be creative and has brought a new and innovative approach to artificial intelligence (AI) study and biology. AI is a unique tool for data and pattern analysis that may be used in a vast range of tasks in general biology, such as enhancing the visual identification of species and monitoring the ecosystems to challenging the prediction of the development of the balance of biological systems [14, 15]. Still, the flash of a change starts from the application of AI in molecular biology, and AI has a wide range of applications. Thankfully, the entry to learn about important molecular structures and functional features has been blazed by the integration of molecular biology and AI [16, 17]. However, it is significant to acknowledge that convolutional neural networks (CNNs) and similar AI algorithms perform rather impressively in analyzing images, neatly outlining the cells and the tissue makeup [18, 19]. Moving further, we see that AI is very effective when identifying gene regulation patterns and metabolic pathways. Sophisticated deep learning methods perform fine-grained analyses of the interdependence of genes, along with unveiling the regulatory networks underlying core cellular functions. At the same time, AI-assisted models map the twisted and complex mechanisms of metabolism, forecast the activity of enzymes, and disentangle the nodes of chemical signaling networks that regulate energy production, cell division, and pathophysiology [20, 21]. AI continues to assert control over multi-omics data analysis and processing, skillfully combining the fragmented aspects of genomics, proteomics, and metabolomics. This work adopts an integrative approach using Random Forests and Bayesian networks to give a complete picture of biological systems and capabilities, affording insight into various unknown relations and emergent phenomena that the reductionist approach would ignore. This perspective sets the stage for predictive biology, looking for biomarkers of disease, mechanistic dissection of new drug targets, and combating the imagination of the patient by mapping the interaction between complex molecular layers. With the entwinement of biology and AI's deeply algorithmic approach, from the macro gene-to-gene view of general biology and positional information to the micro-scale of gene regulation and systems biology, multi-omics integration, and organoids, AI has become an enabling tool in the elucidation of life processes with unprecedented depth and accuracy [22, 23].

2.2 UNDERSTANDING THE BIOLOGY BEHIND SYSTEMS-LEVEL GENE REGULATION

2.2.1 TRANSCRIPTIONAL REGULATION

Buried deep in the laws of respondents' structure, the stage of transcription has always been in the spotlight as the key controller of genetic information. This regulation of genetic activity entails the coordination of several parts of a complex cast, and among this group of players, transcription factors deserve to be noted. These specialized proteins have a distinctive molecular signature that allows the protein to be tightly associated with the DNA sequence or binding site, which is preferred by the transcription factor. These sites can be viewed as "genetic landmarks" on the DNA that are actively involved in controlling the string of translation machinery as it goes through the process of transcribing the DNA code to produce functional RNA molecules [24]. The start of transcription is like a complex and well-rehearsed dance at the molecular level; the transcription factors set the stage for the assembly of the transcription machinery. It is situated within a gene's promoter region and acts as a sort of control on the direction in which transcription will either move forward or stall.

Furthermore, the transcription factors interact not only within the promoter region for genes but also in the enhancer region, which can be present a significant distance away from the gene [25, 26]. This arrangement adds a level of complexity that is akin to that of conducting an orchestra of musicians scattered in various sections, yet the melody is synchronized. What is even more interesting and helpful in understanding their roles is that transcription factors are molecular chameleons that coordinate the process of gene expression. In performing their function, they act as universal switches of gene activity as activators or inhibitors of gene activity as repressors. This binary system doesn't introduce disorder instead, it provides precise and subtle control to enable cells to make highly nuanced adjustments to the environment activating an incredibly complex system of gene expression that is perfectly synchronized with the cell state [27, 28]. Going deeper down to the molecular level of organizational complexity, the motions of transcribing factors with regulatory regions can be perceived as dialogue. Enhancers, characterized by the presence of this, are regulatory sequences that act via interaction with the promoter regions, which are configured like a launch pad for the RNA polymerase machinery to enable precise initiation of transcription. Thus, enhancers interact with promoters using a series of topological transcription factor–mediated chromatin interactions. This molecular conversation implies a coordinated, dynamic regulation of genes through the activities of transcription factors, translating the complex cellular context into the corresponding patterns of gene imprinting. The individual moves involved in transcription and the cooperatively painted transcription factors form a grand masterpiece of regulation. This molecular conversation implies a coordinated, dynamic regulation of genes through the activities of transcription factors, translating complex cellular contexts into precise patterns of gene expression. Much like a symphony, each transcription factor plays its part in harmony, ensuring that gene regulation is both responsive and finely tuned. This intricate orchestration highlights the elegance of molecular mechanisms, revealing a deeply interconnected system that dictates cellular function with remarkable precision [29, 30].

2.2.2 EPIGENETIC REGULATION

In addition to transcription factors, the ever-entertaining story of gene regulation continues its next phase in the sphere of epigenetic modifications. These highly coordinated epigenetic regulations include DNA methylation as well as the changes in histone proteins, which remarkably control the accessibility to and utilization of genes by the transcription machinery. Thus, it is in terms of this epigenetic language that the rich and detailed orchestration of gene expression and all the actualities of the living is encoded [31]. Epigenetic modifications, which are similar to molecular annotations, delicately imprint the genome while also controlling the accessibility of the genes and, hence, whether these are locked for or unlocked for use. One such modification includes DNA methylation, where methyl groups are covalently added to the cytosine base in specific gene regions enzymatically. Contained within this molecular tag is an off switch, so to speak, that carefully regulates which genetic regions are accessible to the transcriptional machinery, ensuring precise control over gene activation. This concept of gene regulation is most active during development and fine-tunes cell employment into differentiated lineages through the mastermind variable gene inactivation [32].

Histone modifications, the other actors in this epigenetic opera, further dress up the proteins around which the DNA is spiraled. The histone tails, which protrude, look almost like the strings of a violin and can be modified depending on the type of chemical groups, like the acetyl or methyl groups. In this case, the example is the acetylation of histones where changes in specific genes signal an "open stage", meaning that the chromatin structure is relaxed to allow transcription factors to access the genes they regulate. On the other hand, histone methylation can serve either a positive or a negative function. It may vary depending on the type of histone that is being methylated and which particular residue is undergoing the process.

Similarly, while some marks may promote gene expression, others may repress it, like a musician who draws fluidity from score changes to evoke contrasting moods. Going into the molecular dynamics, we get introduced to the "writers," "readers," and "erasers" of epigenetic marks. DNA methylation is a comprehensive process catalyzed by DNA methyltransferases that covalently bind methyl groups to cytosine.

Histone modifications are added in an orchestrated manner by a group of enzymes called writers that form part of the histone code that determines chromatin structure and gene regulation. These modifications are then "read" by specific proteins that are known as epigenetic readers, and these readers, in turn, work to interpret the code, recruiting various machinery of the cell. In the same vein, there are also "eraser" enzymes, including histone demethylases, which remove these marks and thus make the genome a reversible process [33]. These epigenetic modifications span across generation alchemy and form a complex web, indeed. This aspect becomes possible due to the ability of these modifications to be passed down through cell divisions while ensuring each cell sustains its respective gene expression patterns. However, it is equally noteworthy how epigenetics is flexible simultaneously because the field directly connects genes and the environment. These molecular tags are improved of an organism, and their annotations depend on physical and chemical conditions like nutrition, stress, and chemicals in vivo that shape the form of the response of the organism to its environment. It is this interplay that forms a living link between the original genome and the score written in epigenetic imprints that is enforced by the environment. It is important to remember that epigenetic regulation carries implications for various domains of biology [34]. The determination of cell fate and tissue morphogenesis requires delicate control of epigenetic marks, including DNA methylation and histone modifications that occur during embryonic development to turn stem cells into specialized cells with distinct functions and structures that make up an organism [35, 36]. On the same note, deviation from these epigenetic patterns is associated with many diseases; the most evident type is DNA methylation, particularly in cancer [37–39].

Nevertheless, epigenetics is plastic in terms of development and environment, and even siblings, who are clones, differ due to the dissimilar experiences they have in their lifetime. Above all, it becomes clear that the enthralling story of how genes are controlled only gets richer as the scientific community continues to discover the fundamental mechanisms behind epigenetic changes. Like a conductor in an orchestra shaping the output of his ensemble, epigenetic writers, erasers, as well as readers, direct the complex movement of genes. It is a molecular dance at the cellular level that programs a canvass of an organism and draws a line at the crossroads of genetic predisposition and environmental influence using DNA methylation and histone modifications.

2.3 POST-TRANSCRIPTIONAL REGULATION

2.3.1 MESSENGER RNA PROCESSING

Gene expression in eukaryotic organisms involves several regulating steps starting from the transcription of DNA into RNA and culminating in the translation of RNA into proteins. Both transcription and translation in eukaryotic organisms occur through several post-transcription events that prepare the initially transcribed pre-messenger RNA (pre-mRNA) for translation into functional RNA [40]. Among these changes, the cap addition on the mRNA molecule remains a basic and highly controlled process. This is an instance that an enzyme adds a 7-methylguanosine nucleotide, or the 5′ cap, to the mRNA. It is for providing mRNA with the ability to pass through the cell machinery and reach multiple phases of its life cycle. One of its major roles is to offer a cover that protects the mRNA with the background of shortening by enzymes known as exonucleases that break down RNA from the two ends. This protection is essential in that it helps keep a pool of mRNA charged with a longer-lasting capability to enhance translation and hence actualize of proteins [41]. Also, the 5′ cap possesses an extremely important function in gene translation—translation initiation.

The cap structure is dotted with specific recognition sites for the eukaryotic initiation factors, which are some of the key proteins that play an essential role in the assembly of the ribosome and other elements of translation initiation. This recognition helps in the process of attachment of the ribosome to the mRNA and accurately positions the start of the ribosome on the codon designated as the start codon for translation. Therefore, some features, including the 5′ cap, enhance translation capacity and fidelity [42]. In addition to the roles stated above the 5′ cap also helps in the proper removal of introns from the pre-mRNA through RNA splicing, where non-coding sequences, or introns, are cleaved off from mRNA, and the coding sequences, or exons, are connected to form mature mRNA [43]. Also of central importance is the 5′ cap that is necessary for the transportation of the mRNA from the nucleus to the cytoplasm in which translation occurs. The 5′ cap is supposedly involved in processes related to gene expression levels and often in cell differentiation in the larger context of gene expression. It takes part in operations that involve the regulation of appropriate mRNA, which is translated into proteins. Further, the 5′ cap is highly mobile concerning transport signal messages within the cell and directly affects the stability and translation rates of mRNA molecules.

In conclusion, the 5′ cap of eukaryotes is more than just a protective cover—it plays a key role in stabilizing the mRNA, helping it navigate through the cell, and ensuring efficient translation. But the 5’ cap doesn’t work alone. At the other end of the mRNA, the poly(A) tail at the 3’ end also plays a crucial role in stability and translation. These two features work together, like bookends, to protect the mRNA and regulate gene expression. Their coordination highlights the intricate molecular system that keeps gene expression precise and efficient in eukaryotic cells [44].

2.3.2 MICRORNAS AND RNA INTERFERENCE

MicroRNAs (miRNAs) play crucial roles in gene expression regulation and occur at the posttranscriptional level. They are relatively short, predominantly within the size of 21–25 nucleotides, and they recognize target sites in the 3′ untranslated regions (UTRs) of target mRNAs. This binding often results in two major outcomes: direct masking or mRNA degradation [45, 46]. The interaction between the miRNA and the target mRNA facilitates the entry of the RNA-induced silencing complex (RISC) that has Argonaute proteins on it. This complex either promotes degradative processing of the mRNA by enhancing its cleavage or translation inhibition by halting the ribosome assembly. The degree of conservation is another aspect that defines the exactness of miRNA target interactions and their role in internalization. The lack of stoichiometric complementarity between the miRNA and the target mRNA makes it possible for a single miRNA to manage multiple target genes in emergent regulatory prose that constitutes intricate signaling pathways. This makes miRNAs able to regulate gene-expressing processes in detail and be involved in many biological processes such as development, differentiation, and immune response, among others. Moreover, the ability to simultaneously target multiple genes has provided profound evidence that dysregulation of miRNA-mediated gene regulation is involved in numerous diseases; therefore, they are of critical importance in molecular biology and have potential application as targets for therapy [47,48]. In other words, regulation of gene expression and metabolic state occurs at the basic level of transcription initiation of epigenetic modifications as well as the rich and complex processing of mRNA molecules. Research on miRNAs and RNA interference introduces additional layers of post-transcriptional regulation, further increasing the complexity of cellular functions. Figure 2.1 provides a visual aid. The copying of genetic information discussed here in detail is called central dogma, with the significant concern on regulatory levels. However, the regulatory levels described in this exploration are not stagnant; rather, they are involved in an interactive and intricate network of genes, as evidenced by the feedback loops highlighted in Figure 2.2. These loops highlight the fact that genes are not just received passively but are rather "self-organized" and reciprocal in the way they regulate each other. Understanding the importance of these formal details is crucial when it comes to embracing AI, a technique that has the potential to decipher, prognosticate, and control such systems to improve current comprehension and implementation in disease diagnosis, drug targeting, and biotechnology. In other words, this chapter can be conceived of as an opening to AI's role in expanding the range of possible outcomes in molecular biology.

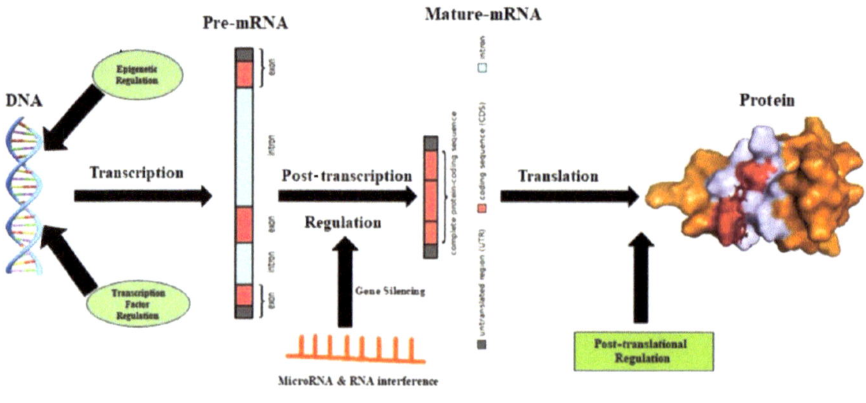

FIGURE 2.1 Gene expression incorporates multiple levels of regulation in the central dogma: Transcriptional and chromatin regulation, as well as microRNA regulation during post-transcription. This complexity highlights the evolving choreography that is being revealed to govern dozens of cellular processes, which could affect future medicine and biotechnology-based applications.

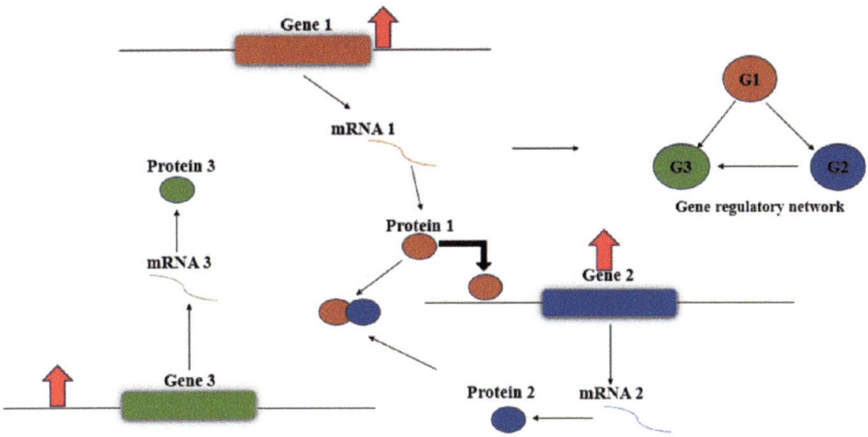

FIGURE 2.2 Proteins 1 and 2 act as transcriptional activators, driving a coordinated gene expression cascade within the regulatory network. The process begins when Protein 1, encoded by Gene 1, triggers a self-regulating loop involving Gene 2 and Protein 2, leading to an increase in both proteins. Together, this dynamic pair enhances the function of Gene 3, resulting in a significant rise in Protein 3 levels. This, in turn, influences enzyme activity and metabolic processes.

2.4 METABOLIC PATHWAYS

The control of metabolic pathways and gene regulation are intimately related, and there is a chance that these two complex biological processes will greatly overlap when it comes to prenatal disorders that result from faulty placenta production during pregnancy. Notably, the regulation of gene expression within these regulatory networks is also highly controlled [49–51]. This regulation is frequently carried out by miRNAs, which can either promote or inhibit genes involved in the metabolic and enzyme functions of metabolites. For instance, the deregulation of miRNA and their target mRNA have been found in preeclampsia or gestational diabetes diseases, which affect genes underlying placental development, vascularization, and hormonal secretion [52, 53]. The following discussion can at least help further the understanding of the pathological processes of chronic diseases and provide potential strategies for their treatment. Thus, leaving behind chronic diseases that actively involve gene regulation together with the pathways of metabolism, there is also a similar connection between human genes and the cholesterol anabolic pathway demonstrated by impaired cardiometabolic and functionally correlated genes, such as *HMGCR* or *SQLE*, intricately expressed in harmony as shown in Figure 2.3, not only in cholesterol synthesis but reflecting a broader phenomenon across diverse biological processes. This association goes as far as glycolysis and cholesterol synthesis and affects the disease processes in conditions like pancreatic cancer, as alterations of gene regulation in this axis are involved in disease phenotypes. This shows that every system interacts with every other one at multiple levels, including gene expression [54]. Pathway

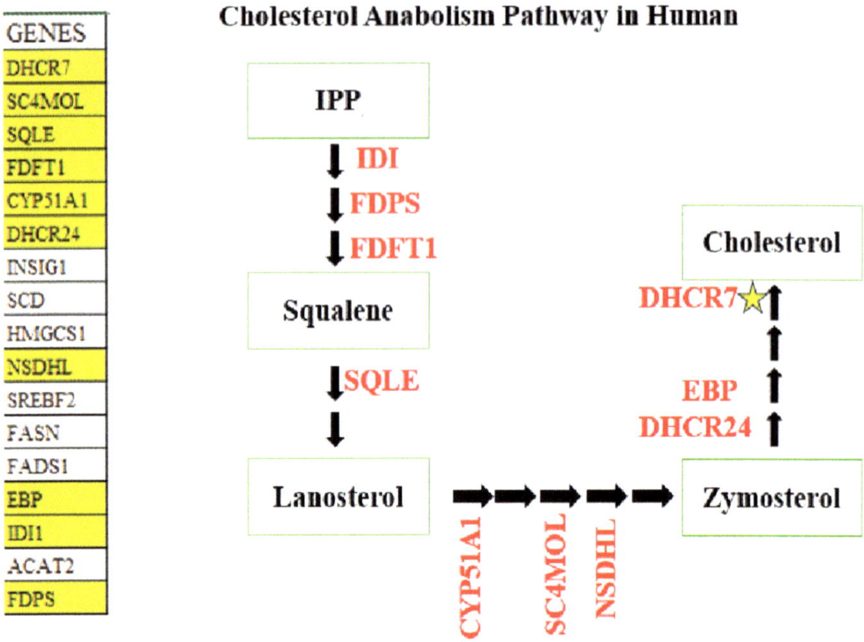

FIGURE 2.3 Functionally correlated genes in a biological pathway are expressed together. The diagram represents the cholesterol anabolism pathway, starting from Acetyl-CoA and proceeding through key intermediates. IPP (Isopentenyl pyrophosphate) is a fundamental isoprenoid precursor involved in sterol biosynthesis.

analysis, as the preferred avenue for untangling the intricacies of differentially expressed genes and proteins, has evolved over decades. Figure 2.4 schematically illustrates the challenges encountered in this evolution, including gaps in knowledge and the significance of the red arrows in mapping these uncertainties. The general aim of pathway analysis is to evaluate the biological systems in terms of integrated networks, implying that issues such as lack of mapping of relations between the smaller pathways and effects of single nucleotide polymorphisms (SNPs) on the pathways are of great concern. Addressing these challenges is critical in moving the focus of the pathway analysis into the next generation where developments in genomics and proteomic platforms can be utilized to increase the specificity, sensitivity, and relative importance of the pathway analysis in deciphering these complex biological systems.

2.4.1 CELLULAR RESPIRATION

Energy metabolism is highly interdependent, and therefore metabolic maps represent them as seamless networks that orchestrate processes of nutrient conversion

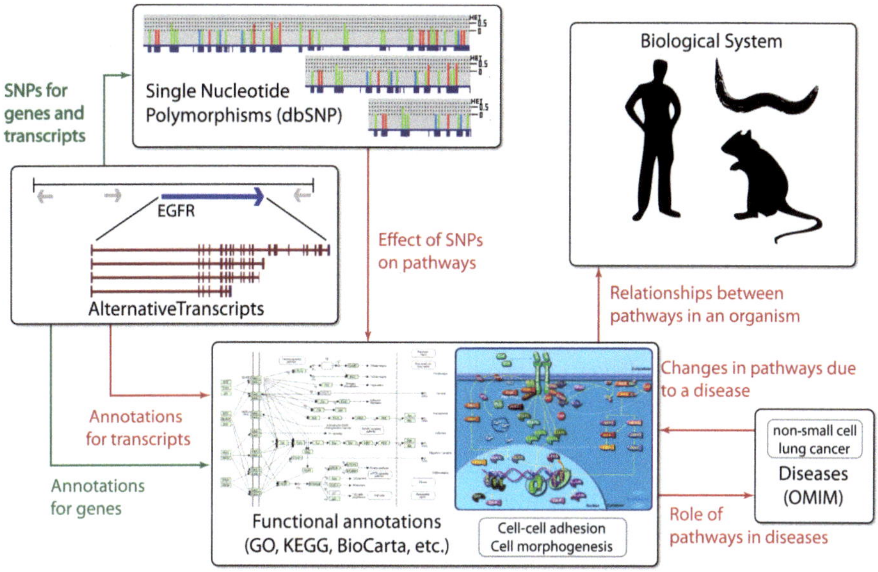

FIGURE 2.4 The knowledge pathway analysis diagram illustrates its development process, highlighting key insights gained and challenges encountered across three discussed generations. The green arrows represent areas with strong and wellsupported information, while the red arrows indicate regions with limited, conflicting, or misleading data. These complexities are particularly important for the next generation of pathway analysis approaches, which aim to integrate technological advancements for a more precise and comprehensive understanding of biological systems. *Abbreviations*: EGFR, Epidermal Growth Factor Receptor; GO, Gene Ontology; KEGG, Kyoto Encyclopedia of Genes and Genomes; BioCarta, BioCarta Pathway Database; OMIM, Online Mendelian Inheritance in Man. [55].

into energy and other essential molecules vital to various cellular processes relevant to cell and organism longevity. An example of tissue-level organization is cellular respiration, a process that happens in cells, preceding glycolysis, and further breaking down glucose to produce adenosine triphosphate (ATP). The energy carrier ATP is involved in the funding of several processes in a cell. Cellular respiration encompasses interconnected stages: Glycolysis begins the breakdown of glucose in the cytoplasm and generates ATP and electron carriers nicotinamide adenine dinucleotide, reduced form (NADH), and the citric acid cycle (Krebs cycle) that occurs in the mitochondria further processes glycolysis products, forming more ATP, NADH, and flavin adenine dinucleotide, reduced form (FADH2). The next process, known as oxidative phosphorylation, occurs in the inner mitochondrial membrane where electron transport chain transfers electrons from NADH and FADH. Oxygen plays a particularly central role as the final electron acceptor in the process of cellular respiration. In combination, metabolic processes, focusing on cellular respiration, harvest energy from nutrients to produce essential molecules and effectively generate ATP for a variety of complex processes essential to life [56]. Metabolic pathways are complex, overly branching networks of chemical processes, which are related to gene regulation because they modulate as well as regulate the process of gene expression. More notably, whereas it has been discussed earlier in the performance of cellular respiration, the regulation of genes is instrumental in controlling the mechanisms in every occurrence of the professed pathway. MicroRNAs, small RNA molecules, are involved in the regulation of gene expression at the post-transcriptional level as they bind to target sequences on mRNAs, thus likely to preventing the translation process or leading to degradation of the mRNAs. In any case, the downregulation of glycolytic and oxidative enzyme expression and upregulation of these miRNA targets would not have directly impaired the organelles' ability to carry out cellular respiration, although it may have affected their rate. It helps to maintain the optimum working of this pathway and keep it responsive to the changes in the cellular environment. For example, it might control glycolytic enzymes by modulating their synthesis rates depending on the energy status of a cell, through miRNAs. Also, it has been found that miRNAs could target the proteins that regulate the use of oxygen for ATP synthesis, altering the aerobic glycolysis-dependent energy metabolism. It has been hypothesized that dysregulation of miRNA targets that are located within metabolic pathways could affect cellular energy homeostasis and this may lead to metabolic disorders [57–59]. Moreover, the link between gene regulation and metabolic pathways is very close and not only applies within a gene framework but also in broader biological contexts including diseases like cancer. This phenomenon also causes the Warburg effect whereby the metabolism of the biological tissue changes, meaning that cancer cells will preferentially conduct glycolysis even if oxygen is available. This transition is also simultaneously associated with alterations in miRNA patterns, which control key metabolic genes and processes [60, 61]. Taking a brief tour of the molecular basis of gene regulation, it has been felt that there exists a similar harmony between transcriptional, epigenetic, and post-transcriptional regulation. Moving from the exactness of transcription factors to the complex regulation by DNA methyltransferases and histone modifiers, the role of each molecular component is comprehensive in the complex tale of gene regulation. Thus, stepping further into the post-transcriptional territory, the focus turns to microRNAs and the members of the translation initiation complex and regulatory proteins. In this information age, where individuals are presented with a plethora of information, it becomes challenging to come up with a conclusive idea about a particular topic or area of interest. Table 2.1 (*Continued*) summarizes these insights, including a wide range of molecular mechanisms, their substrates, and the phenomenal functions of the machinery in orchestrating complicated biological processes. It outlines the current knowledge regarding gene regulation and opens the path to a more detailed examination of the possibility of using artificial intelligence to understand these molecular systems.

TABLE 2.1
Overview of Molecular Machinery Involved in Gene Regulation and Pathway Modulation

Sr No	Molecular Machinery	Targets	Role/Function	Level of Regulation	Reference
1	Transcription factors	Specific DNA (promotors and enhancer)	Gene expression regulation	Transcription	[24]
2	RNA polymerase II	Gene promoters	Transcription initiation and elongation	Transcription	[25, 29]
3	RNA polymerase III	Intragenic A and B block promoter	Synthesis of ribosomal rRNAs, tRNAs, and noncoding RNAs	Transcription	[30]
4	DNase 2	Cytoplasmic accumulation of nuclear DNA	Regulates SASP	Transcription	[27]
5	DNMTs	Methylation at the 5th carbon of cytosine (5mC)	Regulates DNA methylation associated with diverse biological functions	Epigenetic	[31]
6	METTL3 and METTL14	Internal mRNA modification: m6A	Regulates m6A modification Linked to breast and lung cancer progression	Epigenetic	[31]
7	HMTs/HATs/ HDMTs/ HDACs	Histone modification	Modulates chromatin-oriented gene expression	Epigenetic	[32, 33]

TABLE 2.1 (*Continued*)
Overview of Molecular Machinery Involved in Gene Regulation and Pathway Modulation

Sr No	Molecular Machinery	Targets	Role/Function	Level of Regulation	Reference
8	TLR4 ligand LPS	Stimulation of human monocytes	Alternative polyadenylation affecting mRNA stability, localization, and translation efficiency	Post-transcriptional Regulation	[40]
9	snRNPs U1 and U2	5′ and 3′ splice sites of introns	Splicing of pre-mRNA intronic sequences	Post-transcriptional regulation	[40]
10	eIF2	40S ribosomal subunit	Critical for recognizing the start codon and recruiting the 60S ribosomal subunit	Post-transcriptional regulation	[40]
11	PABP	Poly(A) tail located at mRNA 3′ ends	Enhances translation by promoting mRNA pseudo-circularization for improved ribosome recycling	Post-transcriptional regulation	[40]
12	YB-1 (p50)	5′Cap of mRNA	mRNA stabilization	Post-transcriptional regulation	[41]

TABLE 2.1 (*Continued*)
Overview of Molecular Machinery Involved in Gene Regulation and Pathway Modulation

Sr No	Molecular Machinery	Targets	Role/Function	Level of Regulation	Reference
13	PI3K/Akt/mTOR pathway	Phosphorylation of downstream targets	Control of Cap-dependent mRNA translation	Post-transcriptional regulation	[42]
14	eIF4E	5′Cap of mRNA	Initiation of translation	Post-transcriptional regulation	[44]
15	miRNAs	3′Untranslated region of mRNA	Regulates cellular activities via RNA silencing	RNA silencing	[45]
16	RISC	AGO protein family	Complementary RNAs	Regulates gene expression via mRNA cleavage, degradation, or translational repression & RNA silencing	[47]
17	miR-320, miR-123a, miR-422, miR-506, and miR-136	PI3K	PI3K regulation impacting cancer metabolism	Pathway regulation	[57]
18	miR-143	HK2	Directly regulates intracellular glucose levels	Pathway regulation	[57]

TABLE 2.1 (*Continued*)
Overview of Molecular Machinery Involved in Gene Regulation and Pathway Modulation

Sr No	Molecular Machinery	Targets	Role/Function	Level of Regulation	Reference
19	miR-378	PGC-1β, ERRγ pathway	Mediates metabolic shift in breast cancer cells	Pathway regulation	[57]
20	MiR-375	Pancreatic endocrine cell	Insulin regulation	Pathway regulation	[57]

Abbreviations: AGO, Argonaute; DNMTs, DNA methyltransferases; eIF2, eukaryotic initiation factor 2; eIF4E, eukaryotic translation initiation factor 4E; HATs, histone acetyltransferases; HDACs, histone deacetylases; HDMTs, histone demethylases; HK2, hexokinase 2; HMTs, histone methyltransferases; METTL3, methyltransferase 3; METTL14, methyltransferase 14; PABP, poly(A)-binding protein; PI3K, phosphoinositide 3-kinase; RISC, RNA-induced silencing complex; SASP, senescence-associated secretory phenotype; snRNPs, small nuclear ribonucleoproteins; TLR4, toll-like receptor 4.

2.5 AI'S IMPACT ON MOLECULAR BIOLOGY

The relationship between AI and molecular biology has been an interesting paradigm that has helped in the understanding of the mechanism of living. Within this context, AI has utilized its strength to dissect the complexity of gene regulation, metabolic pathways, and systems biology regulatory networks including the integration of multi-omics data.

Deciphering Gene Regulation: These employ AI, primarily convolutional neural networks (CNNs) and recurrrent neural networks (RNNs), and RNNs, that has sequentially enlightened a thorough line of gene and regulatory mechanisms, so to speak, with proficient accuracy. The ability to predict the location of transcription factors, which in turn helps in identifying enhancers and promoters and even understanding how mutation affects gene regulation, brings out the regulation of gene expression as the central platform to the understanding of developmental biology-related diseases and medicine.

Unraveling Metabolic Pathways: The sequences of events in life that involve biochemical reactions that define energy, growth or destruction, and the beginning and progression of diseases have been solved with the use of AI's computer platforms. AI can duplicate many chemical processes, including illnesses like cancer and metabolic disorders, and solve enzyme properties for the pattern to locate the metabolomics central hubs. In addition, through improving substrate selection and pathway optimization, AI can help metabolic engineering strategies to produce sought-after molecule constitutions, from drugs to fuels. It would be relevant to note that integrating genomics, proteomics, and metabolomics data is not easy; therefore, AI, which can manage and integrate large datasets, is extremely beneficial by using Random Forests, Bayesian networks, and deep learning architectures.

Multi-Omics Integration: AI analyzes multi-omics data and unveils a complex gene/protein/metabolite interconnection. Under this system-thinking vision, one arrives at predictive biology, which allows for disease signatures, therapeutic molecular targets, and the anticipatory evaluation of tailored patient reactions to these medicines. The disciplines of artificial intelligence and molecular biology coalesce harmoniously and profoundly, giving birth to a brilliant array of possibilities ranging from epistemological shifts to revolutionary biomedical and biotechnological breakthroughs.

2.5.1 CNNS, SIAMESE NETWORKS, AND THE MOLECULAR CODE

CNNs are particularly well-celebrated for their ability in pattern detection and have transgressed into the realm of genomic data analysis. CNNs were initially invented to recognize various patterns in pixel arrangement displays, and today they are applied to complex ATGC sequences of DNA [61]. This change in perspective echoes in the strings that are intertwined in complex DNA and unveils the revolution in the manner in which the existence of life-forms is perceived. In the same manner, the Siamese neural networks, which were famous for their use in facial recognition, are equally used here to interpret genomic data as represented in Figure 2.5. Based on the information regarding the similarity between faces and Siamese networks, the disclosed relations inherent in DNA sequences explain how various aspects interact with each other on the genetic level [62, 63]. Genomic analysis rides on these networks and opens up tremendous possibilities. One of the great discoveries is the

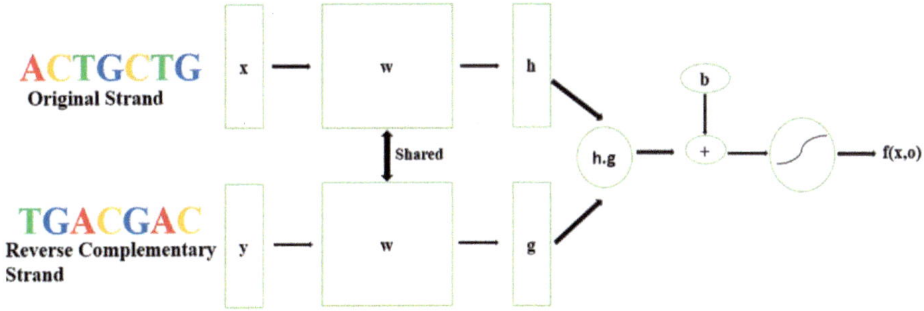

FIGURE 2.5 In genomics, where data identification of some of the most complicated patterns is achieved in addition to finding correlations between different sequences, Siamese networks that found use in text recognition were found to be of great utility. Naturally, this has consequences for improving research on certain genetic traits and molecular diagnostics in the medical field.

power to estimate potentially binding sequences of transcription factors and DNA (TF-DNA). In their capacity as sequels, TFs can regulate the number and speed with which genes are activated and deactivated and, thus, control complex cellular behaviors. This relation is instrumental in deciphering how the genes are switched on or off—a fundamental concept in biological regulation. Whereas TF-DNA binding resembles toggles that decide whether a specific gene is active or inactive, such awareness forms the blueprint for controlling biological activities [64]. By deciphering these interactions, therefore, we go beyond mere comprehension, tackling straight power mastery of life as an activity. Cell regulation and the outcomes of gene expressions such as switching genes on or off or interpreting the results for protein design become tangible, presenting potential for scientific research and medical discoveries.

2.5.2 ChiP-SEQ ANALYSIS WITH DEEP LEARNING

The integration between chromatin immunoprecipitation sequencing (ChIP-seq) analysis and deep learning heralds a new dawn of dissecting complexities in gene regulation and related diseases. Thanks to the use of deep learning models, which introduce a range of quite universal features, the limitations caused by time-consuming ChIP-seq experiments are avoided, which makes the investigation of TF-DNA interactions faster. These models based on the structure of neural networks of deep learning are trained from a vast array of experimental data constructs that include ChIP-seq signal values as well as the associated genomic sequences [65]. The underpinning of the innovation revolves around neural networks and their capacity to recognize intricate features that you might fail to observe. The architecture of the deep learning model is elegantly designed: It takes the sequences of both strands as inputs and gives out predictions for signal values for ChIP-seq, as shown in Figure 2.6. In terms of model development, as the model is fine-tuned on a larger sample of experimental data obtained in successive iterations, the model's ability to make accurate predictions improves. This has provided researchers with a valuable resource that does exactly what they could only dream of before, predicting transcription factor binding affinities at a scale and speed that would have been inconceivable just a short while ago [66]. In addition, this increased dimensionality can create more difficulties in the analysis of the signal than it can give more significant benefits. However, one of the strengths of this approach is that it generalizes the flow of new data very well. In contrast, the amount of data acquired in ChIP-seq experiments is not a significant issue, as the deep learning models built for this experiment can easily handle extensive data input. This capacity to examine vast amounts of genomic data implies an unprecedented potential for discerning complex regulatory features in the genome. There are various usability scenarios of this deep learning analysis of ChIP-seq data: Going beyond binding affinities, gain-of-function researchers can explore the disease relevance of Genetic Variants. The comparison of the number of ChIP-seq signals of different alleles allows a reader to estimate the functional

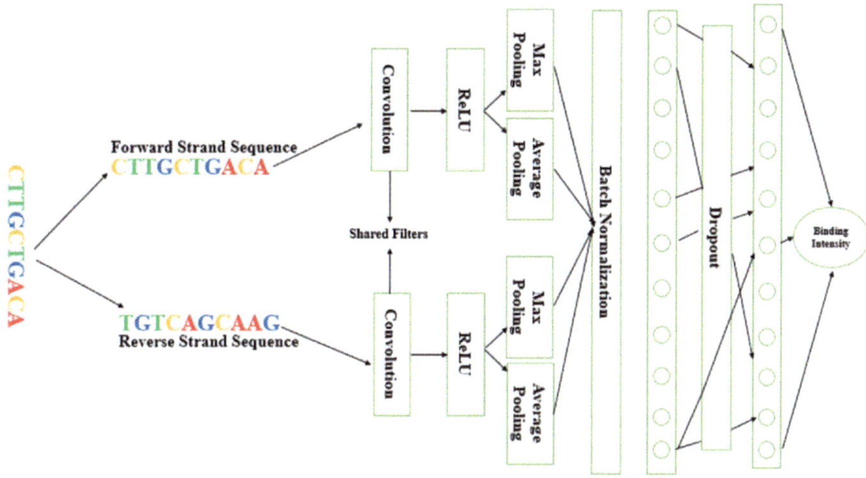

FIGURE 2.6 ChIP-seq neural network for signal prediction. Using information from both DNA strands, this neural network architecture forecasts ChIP-seq signal levels. It utilizes the Rectified Linear Unit (ReLU) activation function, which helps the network introduce non-linearity while preventing vanishing gradients, enhancing model performance. This architecture infinite amount of genomic transcription data and provides enhanced accuracy with more experimental data, all while conserving time.

impact of a variant, as exemplified in the process shown in Figure 2.7. This enables the assessment of variant contribution into phenotypic differences, including such characteristics as hair color or propensity to complex diseases. Altogether, the integration of ChIP-seq analysis and a user-defined deep learning model is a revolutionary advancement in genomics. It overcomes the restraint of experiment restrictions and technical-orientated analysis to boost human accessible areas for research to identify the genetic regulation of the TF-DNA and reveal the pathogenesis of the disease [67,68].

FIGURE 2.7 Understanding complex diseases and evaluating the impact of genetic variations. The method determines whether a variant has a negative effect or is neutral by subtracting the expected ChIP-seq signals for various variants to get functional scores. Information about the genetic makeup of multiple-scaled disorders can be obtained using this method.

2.6 VERSATILE APPLICATIONS OF NEURAL NETWORKS

As we near the end of our journey and delve into strands of genomics and bioinformatics, a distillation of neural network applications is provided in Table 2.2 (*Continued*). This all-embracing resource offers step-by-step guidance on different aspects of many neural network structures, encyclopedic coverage of related tools, myriad forms of input data, categories of output, and practical uses of these structures. From this synthesis, the Siamese neural network demonstrates its competency in sequence processing, and CNNs demonstrate strong predictive capabilities concerning cancer prognostic scenarios. Altogether, these entries demonstrate among population geneticists the versatility and power of such computational models for issues within and beyond genomics involving problems where resolution appears insurmountable without using computational models in fine-tuned tumor diagnostics. In addition to these applications, the responsive interaction between different categories of neural networks across various disciplines reiterates their ability and possible further influences on extant evolving systems. Refer to Table 2.2 (*Continued*) for a distilled overview, providing perspectives and syntheses to work within the explorations of this active and complex area.

2.7 ADVANTAGES AND DISADVANTAGES

Advantage: A new computational framework for studying the molecular biology system that makes use of AI, particularly CNNs and other architectures like artificial neural networks–recurrent neural networks (ANNs-RNNs), for metabolic pathway and gene regulation analysis. When it comes to exceedingly complicated biological information needs—which can be difficult to decipher using the usual experimental approach on this vast network of interactions within a single molecule—artificial intelligence proves to be highly advantageous. CNNs and Siamese networks can tackle pattern recognition analytical problems to reveal data's hidden meanings or robust structural patterns. AI in conjunction with ChIP-seq analysis allows for a detailed examination of cis-regulatory zones and provides insightful information on how these areas control gene activity.

Disadvantage: Nevertheless, there are certain problems with using AI in molecular biology. When using AI models, training deep learning networks is a challenging endeavor that demands a large amount of computing power, even for seasoned data scientists and biologists. The predictions and results generated by some AI models may be less obvious because these algorithms generate "black boxes," which begs the question of how accurate these systems are. Lastly, the fact that biological systems are constantly changing means that there is a continuous process involved that can be quite challenging, as is the unpredictability of developing an AI model that can handle new, unforeseen circumstances. Next, there is an issue of ethics, more specifically, privacy issues connected with the handling of biological data, that should not be disregarded any longer as AI increasingly moves to the foreground of molecular biology as a tool.

TABLE 2.2
Summarized Studies Utilizing Deep Learning in Genomics, Categorized Based on Their Application Focus

Sr No	Architecture	Tool /Pipeline	Input	Output	Application	Reference
1	CNN	Keras	Unstructured gene expression	Classification into tumorous and normal	Cancer prediction	[69]
2	CNN	Keras and tensorflow	Alignment of linked segregating sites with partially shared evolutionary histories	Classification of selective sweeps	Population genetics inference	[61]
3	CNN	Keras and tensorflow	SNP markers	Phenotypic values	Facilitating GWP	[70]
4	Promoter CNN	ALS-Net	ALS patient genomic data	Classification of potential ALS-associated promoter regions	Prediction of ALS phenotype from genotype	[65]
5	CNN	Inception v3, Tensorflow	Histo pathological images	Classify PTCs into BRAFV600E or RAS mutations	Cancer research	[71]
6	Siamese neural network	SENSE	Sequence pair	Sequence comparison	Sequence analysis	[72]

(Continued)

TABLE 2.2 (*Continued*)
Summarized Studies Utilizing Deep Learning in Genomics, Categorized Based on Their Application Focus

Sr No	Architecture	Tool /Pipeline	Input	Output	Application	Reference
7	Siamese neural network	TIMER	DNA sequence	Promotor region	Promotor prediction	[73]
8	Siamese neural network	SemiBin	Contigs	MAGs	Facilitates in metagenomic binning	[62]
9	CNN	DeFine	Raw DNA sequence	Predicts the real-valued ChIP-seq signal value	Predicting transcription factor–DNA interactions	[74]
10	CNN	Tempo MAGE	Mouse midbrain data	Occurrence of H3K27ac within accessible chromatin regions	Predict the activity of enhancers	[75]
11	CNN-RNN	ENNGene	DNA/RNA sequences	Trained models	Facilitates genomic research	[73]
12	CNN	scTIGER	scRNA-seq	GRNs	Predict regulatory mechanisms involved in fundamental cellular processes	[21]

13	Faster R-CNN	PTOLEMI	Biopsy image data	Classify cancer cell as live or dead after drug treatment	Personalized cancer diagnostic	[18]
14	CNN	FPN-SE-ResNet Model	CT scans of abdomens with kidney tumors	Identification of kidney tumors within CT images	Diagnosing and monitoring kidney tumors	[19]
15	DNN	Deep Histone	Peak files of 7 histone modification markers	Modification sites specific to different histone markers	Investigating functional implications of potential disease-linked genetic variants	[66]

Abbreviations: ALS, amyotrophic lateral sclerosis; ChIP-seq, chromatin immunoprecipitation sequencing; CNN, convolutional neural network; CT, computed tomography; GRNs, gene regulatory networks; GWP, genome-wide prediction; PTC, papillary thyroid cancer; SNP, single nucleotide polymorphism.

2.8 CONCLUSION

The study of genes and their rate of synthesis and the organization of metabolic activities is a great adventure into the working of life in organisms. Thus, the conservation of gene expression through different layers, including transcriptional regulation, epigenetic changes, specific post-transcriptional control through miRNAs, and so on, ensures the accuracy as well as flexibility of a cell's reaction to its environment. This chapter has gone outside the conventional experimentation and looked more into the nature that artificial intelligence has brought to the molecular biology study by revealing the hidden features that pattern recognition can uncover. Among the messengers of this change are the CNNs and Siamese networks—two true gems of the deep learning field that are able to decode the molecular level and reveal the structural relationships. More than that, the combination of AI and ChIP-seq analysis has opened the possibility of achieving a comprehensive investigation of the functionality of the cis-regulatory elements and demonstrating the indispensability of these elements in gene regulation. As the code of life unfurls, co-creating with DNA the drama of the world, it seems that understanding how genes turn the pages and how cells read the score of life are not only the answers to a universal riddle but also the potential pathway to revolutionary discoveries in medicine and biotechnology.

REFERENCES

1. Keith A Spriggs, Martin Bushell, and Anne E Willis. Translational regulation of gene expression during conditions of cell stress. *Molecular Cell*, 40(2):228–237, 2010.
2. Alan Wise. Transcriptional switches in the control of macronutrient metabolism. *Nutrition Reviews*, 66(6):321–325, 2008.
3. Michael Levine and Robert Tjian. Transcription regulation and animal diversity. *Nature*, 424(6945):147–151, 2003.
4. Niels de Jonge and Diana B Peckys. Live cell electron microscopy is probably impossible. *ACS Nano*, 10(10):9061–9063, 2016.
5. Talley J Lambert and Jennifer C Waters. Navigating challenges in the application of superresolution microscopy. *Journal of Cell Biology*, 216(1):53–63, 2017.
6. David H Tse, Nicole A Becker, Robert T Young, Wilma K Olson, Justin P Peters, Tanya L Schwab, Karl J Clark, and L James Maher III. Designed architectural proteins that tune DNA looping in bacteria. *Nucleic Acids Research*, 49(18):10382–10396, 2021.
7. Wilma K Olson, Andrey A Gorin, Xiang-Jun Lu, Lynette M Hock, and Victor B Zhurkin. DNA sequence-dependent deformability deduced from protein–DNA crystal complexes. *Proceedings of the National Academy of Sciences*, 95(19):11163–11168, 1998.
8. Ri Han, Hongryul Yoon, Gahee Kim, Hyundo Lee, and Yoonji Lee. Revolutionizing medicinal chemistry: the application of artificial intelligence (AI) in early drug discovery. *Pharmaceuticals*, 16(9):1259, 2023.
9. Monika Khandelwal, Nazir Shabbir, and Saiyed Umer. Extraction of sequence-based features for prediction of methylation sites in protein sequences. In *Artificial Intelligence Technologies for Computational Biology*, pages 29–46. CRC Press, 2022.
10. Ulrich AK Betz, Loukik Arora, Reem A Assal, Hatylas Azevedo, Jeremy Baldwin, Michael S Becker, Stefan Bostock, Vinton Cheng, Tobias Egle, Nicola Ferrari, et al.

Game changers in science and technology-now and beyond. *Technological Forecasting and Social Change*, 193:122588, 2023.

11. Monika Khandelwal and Ranjeet Kumar Rout. PRMxAI: protein arginine methylation sites prediction based on amino acid spatial distribution using explainable artificial intelligence. *BMC Bioinformatics*, 24(1):376, 2023.
12. Valeria D'Argenio. The high-throughput analyses era: are we ready for the data struggle? *High-throughput*, 7(1):8, 2018.
13. Fumimasa Nomura, Takehiko Inaba, Satoshi Ishikawa, Miki Nagata, Sho Takahashi, Hirokazu Hotani, and Kingo Takiguchi. Microscopic observations reveal that fusogenic peptides induce liposome shrinkage prior to membrane fusion. *Proceedings of the National Academy of Sciences*, 101(10):3420–3425, 2004.
14. Patrik Lauha, Panu Somervuo, Petteri Lehikoinen, Lisa Geres, Tobias Richter, Sebastian Seibold, and Otso Ovaskainen. Domain-specific neural networks improve automated bird sound recognition already with small amount of local data. *Methods in Ecology and Evolution*, 13(12):2799–2810, 2022.
15. José Augusto Correa Martins, José Marcato Junior, Marlene Patzig, Diego André Sant'Ana, Hemerson Pistori, Veraldo Liesenberg, and Anette Eltner. Identifying plant species in kettle holes using UAV images and deep learning techniques. *Remote Sensing in Ecology and Conservation*, 9(1):1–16, 2023.
16. Keith T Butler, Daniel W Davies, Hugh Cartwright, Olexandr Isayev, and Aron Walsh. Machine learning for molecular and materials science. *Nature*, 559(7715):547–555, 2018.
17. Arlindo L Oliveira. Biotechnology, big data and artificial intelligence. *Biotechnology Journal*, 14(8):1800613, 2019.
18. Bernard Moerdler, Matan Krasner, Elazar Orenbuch, Avi Grad, Benjamin Friedman, Eliezer Graber, Efrat Barbiro-Michaely, and Doron Gerber. PTOLEMI: personalized cancer treatment through machine learning-enabled image analysis of microfluidic assays. *Diagnostics*, 13(19):3075, 2023.
19. Abubaker Abdelrahman and Serestina Viriri. FPN-SE-ResNet model for accurate diagnosis of kidney tumors using CT images. *Applied Sciences*, 13(17):9802, 2023.
20. Fengjun Zhang, Mingyue Xia, Jiarong Jiang, Shuai Wang, Qiong Zhao, Cheng Yu, Jinzhen Yu, Dexian Xian, Xiao Li, Lin Zhang, et al. Machine learning and bioinformatics to identify 8 autophagy-related biomarkers and construct gene regulatory networks in dilated cardiomyopathy. *Scientific Reports*, 12(1):15030, 2022.
21. Madison Dautle, Shaoqiang Zhang, and Yong Chen. scTIGER: A deep-learning method for inferring gene regulatory networks from case versus control scRNA-seq datasets. *International Journal of Molecular Sciences*, 24(17):13339, 2023.
22. Javier E Flores, Daniel M Claborne, Zachary D Weller, Bobbie-Jo M Webb-Robertson, Katrina M Waters, and Lisa M Bramer. Missing data in multi-omics integration: recent advances through artificial intelligence. *Frontiers in Artificial Intelligence*, 6:1098308, 2023.
23. Xiujing He, Xiaowei Liu, Fengli Zuo, Hubing Shi, and Jing Jing. Artificial intelligence-based multi-omics analysis fuels cancer precision medicine. In *Seminars in Cancer Biology*, volume 88, pages 187–200. Elsevier, 2023.
24. Samuel A Lambert, Arttu Jolma, Laura F Campitelli, Pratyush K Das, Yimeng Yin, Mihai Albu, Xiaoting Chen, Jussi Taipale, Timothy R Hughes, and Matthew T Weirauch. The human transcription factors. *Cell*, 172(4):650–665, 2018.

25. Patrick Cramer. Organization and regulation of gene transcription. *Nature*, 573(7772):45–54, 2019.
26. Ranjeet Kumar Rout, Saiyed Umer, Monika Khandelwal, Smitarani Pati, Saurav Mallik, Bunil Kumar Balabantaray, and Hong Qin. Identification of discriminant features from stationary pattern of nucleotide bases and their application to essential gene classification. *Frontiers in Genetics*, 14:1154120, 2023.
27. Akiko Takahashi, Tze Mun Loo, Ryo Okada, Fumitaka Kamachi, Yoshihiro Watanabe, Masahiro Wakita, Sugiko Watanabe, Shimpei Kawamoto, Kenichi Miyata, Glen N Barber, et al. Downregulation of cytoplasmic DNases is implicated in cytoplasmic DNA accumulation and SASP in senescent cells. *Nature Communications*, 9(1):1249, 2018.
28. Hiroyuki Hosokawa and Ellen V Rothenberg. How transcription factors drive choice of the T cell fate. *Nature Reviews Immunology*, 21(3):162–176, 2021.
29. Allison C Schier and Dylan J Taatjes. Structure and mechanism of the RNA polymerase II transcription machinery. *Genes & Development*, 34(7-8):465–488, 2020.
30. Ian M Willis and Robyn D Moir. Signaling to and from the RNA polymerase III transcription and processing machinery. *Annual Review of Biochemistry*, 87:75–100, 2018.
31. Lin-Yong Zhao, Jinghui Song, Yibin Liu, Chun-Xiao Song, and Chengqi Yi. Mapping the epigenetic modifications of DNA and RNA. *Protein & Cell*, 11(11):792–808, 2020.
32. Chongyuan Luo, Petra Hajkova, and Joseph R Ecker. Dynamic DNA methylation: in the right place at the right time. *Science*, 361(6409):1336–1340, 2018.
33. Bruce Stillman. Histone modifications: insights into their influence on gene expression. *Cell*, 175(1):6–9, 2018.
34. Céline Tiffon. The impact of nutrition and environmental epigenetics on human health and disease. *International Journal of Molecular Sciences*, 19(11):3425, 2018.
35. Ozren Bogdanović and Ryan Lister. DNA methylation and the preservation of cell identity. *Current Opinion in Genetics & Development*, 46:9–14, 2017.
36. Ruimin Xu, Chong Li, Xiaoyu Liu, and Shaorong Gao. Insights into epigenetic patterns in mammalian early embryos. *Protein & Cell*, 12(1):7–28, 2021.
37. Alexander Koch, Sophie C Joosten, Zheng Feng, Tim C de Ruijter, Muriel X Draht, Veerle Melotte, Kim M Smits, Jurgen Veeck, James G Herman, Leander Van Neste, et al. Analysis of DNA methylation in cancer: location revisited. *Nature Reviews Clinical Oncology*, 15(7):459–466, 2018.
38. Yunbao Pan, Guohong Liu, Fuling Zhou, Bojin Su, and Yirong Li. DNA methylation profiles in cancer diagnosis and therapeutics. *Clinical and Experimental Medicine*, 18:1–14, 2018.
39. Abeer M Mahmoud and Mohamed M Ali. Methyl donor micronutrients that modify DNA methylation and cancer outcome. *Nutrients*, 11(3):608, 2019.
40. Susan Carpenter, Emiliano P Ricci, Blandine C Mercier, Melissa J Moore, and Katherine A Fitzgerald. Post-transcriptional regulation of gene expression in innate immunity. *Nature Reviews Immunology*, 14(6):361–376, 2014.
41. Valentina Evdokimova, Peter Ruzanov, Hiroaki Imataka, Brian Raught, Yuri Svitkin, Lev P Ovchinnikov, and Nahum Sonenberg. The major mRNA-associated protein YB-1 is a potent 5' cap-dependent mRNA stabilizer. *The EMBO Journal*, 20(19):5491–5502, 2001.
42. Y Mamane, E Petroulakis, O LeBacquer, and N Sonenberg. mTOR, translation initiation and cancer. *Oncogene*, 25(48):6416–6422, 2006.
43. Mutsuhito Ohno, Hiroshi Sakamoto, and Yoshiro Shimura. Preferential excision of the 5' proximal intron from mRNA precursors with two introns as mediated by the cap structure. *Proceedings of the National Academy of Sciences*, 84(15):5187–5191, 1987.

44. Joe D Lewis and Elisa Izaurflde. The role of the cap structure in RNA processing and nuclear export. *European Journal of Biochemistry*, 247(2):461–469, 1997.
45. Richard W Carthew. Gene regulation by micrornas. *Current Opinion in Genetics & Development*, 16(2):203–208, 2006.
46. Ramesh A Shivdasani. Micrornas: regulators of gene expression and cell differentiation. *Blood*, 108(12):3646–3653, 2006.
47. Hiro-oki Iwakawa and Yukihide Tomari. Life of RISC: formation, action, and degradation of RNA-induced silencing complex. *Molecular Cell*, 82(1):30–43, 2022.
48. Mengfan Pu, Jing Chen, Zhouteng Tao, Lingling Miao, Xinming Qi, Yizheng Wang, and Jin Ren. Regulatory network of miRNA on its target: coordination between transcriptional and post-transcriptional regulation of gene expression. *Cellular and Molecular Life Sciences*, 76:441–451, 2019.
49. Bing Chen, Hongbin Li, Xiao Zeng, Pengbo Yang, Xinyu Liu, Xia Zhao, and Shufang Liang. Roles of microRNA on cancer cell metabolism. *Journal of Translational Medicine*, 10(1):1–12, 2012.
50. Tong Mou, Di Zhu, Xufu Wei, Tingting Li, Daofeng Zheng, Junliang Pu, Zhen Guo, and Zhongjun Wu. Identification and interaction analysis of key genes and micrornas in hepatocellular carcinoma by bioinformatics analysis. *World Journal of Surgical Oncology*, 15(1):1–9, 2017.
51. Sugarniya Subramaniam, Varinder Jeet, Judith A Clements, Jennifer H Gunter, and Jyotsna Batra. Emergence of micrornas as key players in cancer cell metabolism. *Clinical Chemistry*, 65(9):1090–1101, 2019.
52. Guodong Fu, Jelena Brkić, Heyam Hayder, and Chun Peng. Micrornas in human placental development and pregnancy complications. *International Journal of Molecular Sciences*, 14(3):5519–5544, 2013.
53. Asghar Ali, Frieder Hadlich, Muhammad W Abbas, Muhammad A Iqbal, Dawit Tesfaye, Gerrit J Bouma, Quinton A Winger, and Siriluck Ponsuksili. MicroRNA-mRNA networks in pregnancy complications: a comprehensive downstream analysis of potential biomarkers. *International Journal of Molecular Sciences*, 22(5):2313, 2021.
54. Joanna M Karasinska, James T Topham, Steve E Kalloger, Gun Ho Jang, Robert E Denroche, Luka Culibrk, Laura M Williamson, Hui-Li Wong, Michael KC Lee, Grainne M O'Kane, et al. Altered gene expression along the glycolysis–cholesterol synthesis axis is associated with outcome in pancreatic cancer. *Clinical Cancer Research*, 26(1):135–146, 2020.
55. Purvesh Khatri, Marina Sirota, and Atul J Butte. Ten years of pathway analysis: current approaches and outstanding challenges. *PLoS Computational Biology*, 8(2):e1002375, 2012.
56. Sarah Stauffer, Aaron Gardner, Dewi Ayu Kencana Ungu, Ainara López-Córdoba, and Matthias Heim. *Labster Virtual Lab Experiments: Basic Biology*. Springer, 2018.
57. Z Chen, Y Li, H Zhang, P Huang, and R Luthra. Hypoxia-regulated microRNA-210 modulates mitochondrial function and decreases ISCU and COX10 expression. *Oncogene*, 29(30):4362–4368, 2010.
58. Lillian J Eichner, Marie-Claude Perry, Catherine R Dufour, Nicholas Bertos, Morag Park, Julie St-Pierre, and Vincent Giguère. miR-378* mediates metabolic shift in breast cancer cells via the PGC-1β/ERRγ transcriptional pathway. *Cell Metabolism*, 12(4):352–361, 2010.
59. Na Li, David J Bates, Jin An, Derek A Terry, and Eugenia Wang. Up-regulation of key micrornas, and inverse down-regulation of their predicted oxidative phosphorylation target genes, during aging in mouse brain. *Neurobiology of Aging*, 32(5):944–955, 2011.

60. Guomin Shen, Xiaobo Li, Yong-feng Jia, Gary A Piazza, and Yaguang Xi. Hypoxia-regulated micrornas in human cancer. *Acta Pharmacologica Sinica*, 34(3):336–341, 2013.
61. Lex Flagel, Yaniv Brandvain, and Daniel R Schrider. The unreasonable effectiveness of convolutional neural networks in population genetic inference. *Molecular Biology and Evolution*, 36(2):220–238, 2019.
62. Shaojun Pan, Chengkai Zhu, Xing-Ming Zhao, and Luis Pedro Coelho. A deep siamese neural network improves metagenome-assembled genomes in microbiome datasets across different environments. *Nature Communications*, 13(1):2326, 2022.
63. Ediem Al-Jibury, James WD King, Ya Guo, Boris Lenhard, Amanda G Fisher, Matthias Merkenschlager, and Daniel Rueckert. A deep learning method for replicate-based analysis of chromosome conformation contacts using siamese neural networks. *Nature Communications*, 14(1):5007, 2023.
64. Sohyun Bang, Mary Galli, Peter A Crisp, Andrea Gallavotti, and Robert J Schmitz. Identifying transcription factor–DNA interactions using machine learning. *in silico Plants*, 4(2):diac014, 2022.
65. Xiaojiang Xu, Stephen Hoang, Marty W Mayo, and Stefan Bekiranov. Application of machine learning methods to histone methylation ChIP-Seq data reveals H4R3me2 globally represses gene expression. *BMC Bioinformatics*, 11:1–20, 2010.
66. Qijin Yin, Mengmeng Wu, Qiao Liu, Hairong Lv, and Rui Jiang. DeepHistone: a deep learning approach to predicting histone modifications. *BMC Genomics*, 20(2):11–23, 2019.
67. Etienne Routhier and Julien Mozziconacci. Genomics enters the deep learning era. *PeerJ*, 10:e13613, 2022.
68. Zhongxiao Li, Elva Gao, Juexiao Zhou, Wenkai Han, Xiaopeng Xu, and Xin Gao. Applications of deep learning in understanding gene regulation. *Cell Reports Methods*, 3(1):100384, 2023.
69. Milad Mostavi, Yu-Chiao Chiu, Yufei Huang, and Yidong Chen. Convolutional neural network models for cancer type prediction based on gene expression. *BMC Medical Genomics*, 13:1–13, 2020.
70. Patrik Waldmann, Christina Pfeiffer, and Gábor Mészáros. Sparse convolutional neural networks for genome-wide prediction. *Frontiers in Genetics*, 11:25, 2020.
71. Peiling Tsou and Chang-Jiun Wu. Mapping driver mutations to histopathological subtypes in papillary thyroid carcinoma: applying a deep convolutional neural network. *Journal of Clinical Medicine*, 8(10):1675, 2019.
72. Wei Zheng, Le Yang, Robert J Genco, Jean Wactawski-Wende, Michael Buck, and Yijun Sun. SENSE: siamese neural network for sequence embedding and alignment-free comparison. *Bioinformatics*, 35(11):1820–1828, 2019.
73. Eliška Chalupová, Ondřej Vaculík, Jakub Poláček, Filip Jozefov, Tomáš Majtner, and Panagiotis Alexiou. ENNGene: an easy neural network model building tool for genomics. *BMC Genomics*, 23(1):1–11, 2022.
74. Meng Wang, Cheng Tai, Weinan E, and Liping Wei. DeFine: deep convolutional neural networks accurately quantify intensities of transcription factor-DNA binding and facilitate evaluation of functional non-coding variants. *Nucleic Acids Research*, 46(11):e69, 2018.
75. Mohammad Hallal, Mariette Awad, and Pierre Khoueiry. TempoMAGE: a deep learning framework that exploits the causal dependency between time-series data to predict histone marks in open chromatin regions at time-points with missing ChIP-seq datasets. *Bioinformatics*, 37(23):4336–4342, 2021.

3 Assessment of Machine Learning Algorithms in DNA Sequence Data Mining

A. Pathanjali Sastri, Shaik Akbar, and K.Suresh

3.1 INTRODUCTION

The progress in scientific research has enabled us to delve into the riddles of life's complexities through the study of genomics. Over the past few decades, there has been a rapid expansion of biological data, giving rise to extensive databases. From this vast amount of data, valuable insights can be gained, leading to the birth of bioinformatics. Bioinformatics is an interdisciplinary field that combines mathematics, life sciences, and computer science to extract meaningful information from biological data. It serves as a guide for researchers in their biological investigations. The first step involves extracting information from genomic DNA sequences to identify protein-coding regions. Next, researchers use simulations and predictions to determine the spatial structure of proteins. Finally, based on the function of these proteins, researchers can design necessary drugs. The amount of organic information is exponentially increasing every 18 months, according to research. In 1982, GenBank nucleic acid sequence database contained only 606 sequences and 680,000 nucleotide bases. However, as of February 2013, the database now holds a staggering One hundred sixty-two million genetic sequence records with approximately One hundred and fifty billion nucleotide bases. Bioinformatics is crucial in mining knowledge from this vast data and guiding biological research [1] [2]. Dealing with complex biological information involves two main challenges: effectively storing and managing large amounts of data while extracting meaningful insights. Machine learning is an essential technique for achieving artificial intelligence in this context.

3.2 FUNDAMENTAL UNDERSTANDING OF DNA

HiSeq X Ten [3] is currently the leading sequencing platform in the field of life sciences. The introduction and commercialization of this technology have greatly

DOI: 10.1201/9781032660714-3

advanced the progress of the sequencing industry. With its high throughput and cost-effectiveness, HiSeq X Ten [3] has democratized sequencing, making it more accessible than ever before. Sequence similarity plays a crucial role in sequence data mining and is highly valued in bioinformatics. It measures the likeness between sequences, with a similarity exceeding 30% suggesting potential homology. Homologous sequences share a common evolutionary lineage and may exhibit similarities in their structures and functions.

3.3 ADVANCEMENT IN SEQUENCING METHODOLOGIES

The field of biological information technology has witnessed significant advancements in sequencing techniques over the years. These developments can be categorized into three distinct phases. The first phase, known as the first-generation sequencing technology, involved methods such as the chain termination method by Sanger [4] and the chain degradation method by Gilbert [5]. While Sanger sequencing still finds utility in conventional sequencing and validation, its high cost and low throughput have limited its widespread adoption on a large scale [4].

However, over the course of more than four decades, sequencing technology has made remarkable progress. Newer techniques and methodologies have emerged, offering enhanced efficiency and affordability. These advancements have enabled researchers to overcome many of the limitations associated with early sequencing methods. Today, with state-of-the-art sequencing technologies at our disposal, scientists are able to unlock crucial insights into various biological processes with unprecedented ease and accuracy. The continuous evolution of sequencing technology holds great promise for further discoveries In the realm of biological investigation.

The introduction of Sequencing technology of the second generation, spearheaded by the 454 sequencing (developed in 2005), marked a significant breakthrough after relentless research efforts. These systems allowed for the simultaneous analysis of billions of sequencing reactions, revolutionizing the field with its high-speed and cost-effective capabilities. Today, second-generation sequencing reigns supreme in scientific research due to its widespread adoption and affordability.

In 2011, third-generation sequencing took center stage with Oxford PacBio's Single Molecule Real-Time (SMRT) technology enables the sequencing of individual molecules leading the way. The flagship feature of this generation is single molecule sequencing, but continuous adjustments and upgrades are necessary for its large-scale application. These advances in sequencing technology are transforming enabling personalized medicine through diagnostic sequencing capabilities offered via high-throughput options. Figure 3.1 presents an example illustrating the history of nucleotide sequence analysis.

The field of biomedicine is now focused on genomic big data analysis, which encompasses database management, sharing, interpretation and quality assurance. It is crucial to understand the specific platform used for sequencing as each platform has distinct error rate profiles, biases, limitations that affect data generation. To support biological data analysis, Leading global repositories of biological data such as DDBJ, NCBI, EBI and DDBJ have established comprehensive

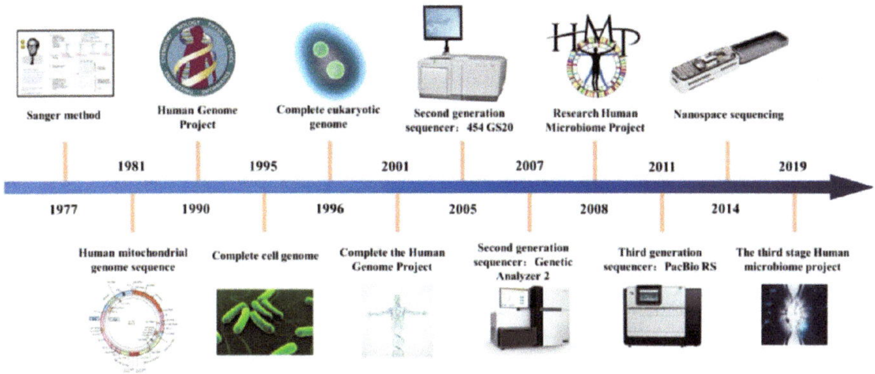

FIGURE 3.1 History of nucleotide sequence analysis (`http://surl.li/rfoll`).

databases and diverse services. However, biomedical data presents challenges due to its diversity, complexity, rich content with fragmented distribution within high-dimensional frameworks requiring multi-level cross-sharing.

3.4 THE STRUCTURAL ARRANGEMENT OF DNA SEQUENCES

In the field of biology, extensive research has indicated that biological sequences exhibit a structured and organized pattern rather than being random strings. Particularly in the case of DNA sequences, they showcase A sequential configuration of smaller elements. These elements consist of four distinct deoxyribonucleotides (bases), and their specific order holds great significance in influencing the diversity observed in DNA molecules. Figure 3.2 illustrates the arrangement of the DNA double helix. Complementary base pairing is a fundamental phenomenon in the DNA

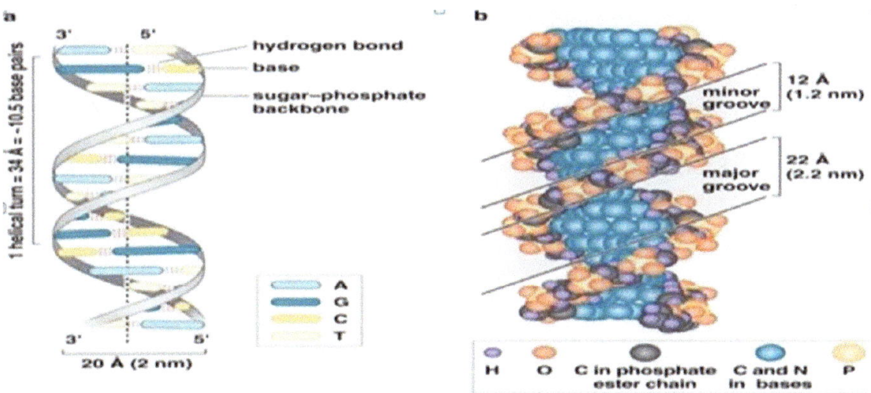

FIGURE 3.2 Structure of double helix DNA.

double helix structure. The nitrogenous bases on one strand selectively Form connections with particular bases on the opposite Genetic thread. This bonding creates base pairs, which are the building blocks of the DNA sequence.

3.5 DNA SEQUENCE DATA IS UNIQUE AND DIFFERENT FROM OTHER TYPES OF DATA IN SEVERAL WAYS

1. It is composed of non-numeric characters, specifically A, T, C, and G, which serve as the fundamental building blocks of the data.
2. The length of DNA sequences can vary significantly, ranging from just a few dozen characters to sizes that extend up to hundreds of megabytes
3. The sequences themselves carry specific biological significance that is intrinsic to their structure and composition.
4. Before any analysis can be performed on DNA sequence data, it requires specific preprocessing steps to correct errors inherent in the Genome sequencing method and eliminate Anomaly present within the data. These Essential steps ensure accurate and reliable results are obtained.

3.5.1 ENCODING OF DNA SEQUENCES

DNA sequence coding refers to the process by which genetic information is stored in DNA molecules. The sequence of nucleotides, which are the building blocks of DNA, determines the specific characteristics and functions of an organism. This coding system allows for the replication and transmission of genetic material through generations. By studying and decoding these sequences, scientists can gain valuable insights into the structure and function of genes, as well as their potential implications for various biological processes. The field of DNA sequence coding plays a crucial role in advancing our understanding of genetics and has numerous applications in areas such as medicine, agriculture, and forensic science.

3.5.2 SIMILARITY AMONG DNA SEQUENCES

The primary mining techniques in machine learning involve data characterization, association, classification, regression, clustering, and outlier analysis. When applied to DNA sequence mining, these approaches are closely tied to sequence similarity analysis [6]. Sequence similarity plays a crucial role in DNA sequence data mining.

Sequence similarity refers to shared or identical segments between sequences, either quantitatively or descriptively. A similarity greater than 30% indicates a homologous relationship between sequences, suggesting a common evolutionary origin. Detecting similarities between an unknown sequence and those with known functions helps predict the function of the unknown sequence [7]. Identifying sequences that surpass a specified similarity threshold is an essential aspect of DNA sequence similarity research. One commonly used method involves constructing a similarity matrix [8] and identifying the best match among sequences while considering insertions, deletions, and mutations. Distinctions exist between global and local similarity research. Global alignment matches the Full sequence and is Adequate for Nucleotide sequences with Maximum overall Homogeneity. The Needleman-Wunsch algorithm [9] exemplifies global alignment [10]. However, given that only about 2% of DNA sequences account for functional genes, studying local similarity proves more meaningful. Subsequence comparison algorithms such as BLAST, FASTA and Smith-Waterman efficiently address this by enabling efficient database similarity searches.

Newer investigations, such as the work of Delibas and Arslan [11], have brought forth groundbreaking methodologies like non-aligned sequence similarity analysis. They utilize texture image similarity calculations, a technique stemming from digital image processing, to revolutionize DNA sequence analysis. Sequence similarity analysis stands as a bedrock in computational biology and bioinformatics, playing a pivotal role in deciphering gene functions, predicting protein structures, and retrieving sequences [12]. The adaptation and choice of appropriate similarity analysis methods aligned with the biological context continue to be imperative in DNA sequence data mining, serving as an indispensable pillar in this domain.

3.6 MACHINE LEARNING ALGORITHM

Over the past few decades, biomedical research and biotechnology have undergone revolutionary advancements, resulting in an explosion of biomedical data. The focus has shifted from accumulating this data to extracting meaningful insights from it. In one aspect, the quick progress in Researching biotechnology and conducting analyses on biological data has given rise to a Intimidating yet promising Domain: bioinformatics. Conversely, the continual evolution of Genomic data mining techniques has yielded numerous efficient and scalable algorithms. The critical task now is to bridge machine learning and bioinformatics effectively, enabling the successful analysis of biomedical data.

It is crucial to explore how data mining can be harnessed for impactful biomedical data analysis. Identifying ways to integrate these fields could lead to the Advancement of potent Bioinformatics-driven machine learning approaches. This involves analyzing how machine learning methods can be optimized for Efficient analysis of biomedical data and outlining key research queries that can drive the further enhancement of these algorithms [13].

3.7 EXPLORING THE CORE STAGES OF DATA MINING: A FUNDAMENTAL OVERVIEW

Data mining is an interdisciplinary field that amalgamates traditional statistical methods with algorithms from computer science. Its primary objective is to extract valuable insights from extensive datasets for scientific, computational or industrial applications. We provide a comprehensive delineation of the data mining process across six key facets as illustrated in Figure 3.3.

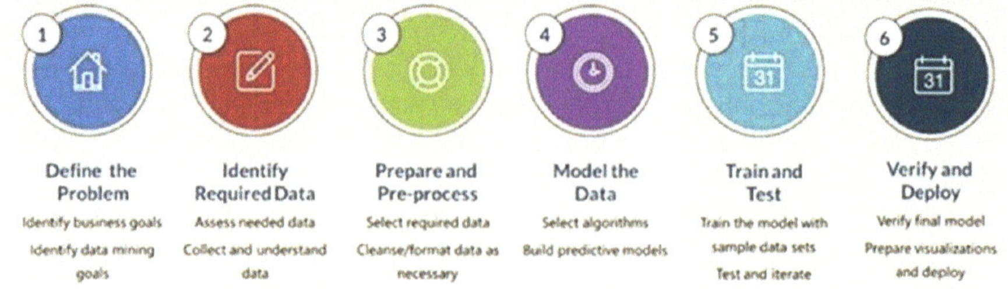

FIGURE 3.3 Core stages of data mining.

3.7.1 AFFILIATION (OR ASSOCIATION) RULE MINING ALGORITHM

Affiliation rule mining is a pivotal feature of information mining, diving into the ID of affiliations and incessant examples inside a given dataset. The procedure involves dealing with two main subproblems: 1) establishing a minimum support threshold criterion to reveal association rules that adhere to defined limitations on these high-support items. 2) extracting High-support itemsets from the database using a minimum support count. Association rule mining has been successful in various fields, including medical data examination and virtual shopping basket analysis, in addition to its significance in studying business data. The Apriori calculation stands apart as a prototypical affiliation rule-based mining calculation, exhibiting its flexibility in applications going from grouping design mining to protein structure forecast. Determined calculations in information mining frequently track down motivation from Apriori Zhang et al. [14]. The essential way to deal with affiliation rule mining is utilizing different measurements to investigate the strong relationship inside the information base. Among these measurements, the least help and least certainty are generally used. The Apriori calculation embraces a directed procedure to uncover affiliation rules between information things inside the data set.

3.7.2 CLASSIFICATION ALGORITHM

Machine learning extensively investigates classification, a foundational task in the field. At its core, classification leverages a Dependent variable to foresee the class of the Dependent variable designated by the user. The primary focal points in genetic analysis lie in genome classification and sequence annotation. Various algorithms encompassing fuzzy sets are prevalent in mining genome sequences: genetic algorithms, neural networks and rough sets. Additionally, a host of overarching classification models exists, including but not limited to naive Bayesian networks, Decision Trees (DT), Neural Networks (NN), and rule learning through evolutionary algorithms.

3.7.3 CLUSTER ANALYSIS TECHNIQUES

The Clustering algorithms serve the purpose of grouping sequences with similar characteristics, allowing the extraction of valuable information from known functions and structures in biological sequences. This clustering process holds significant importance in advancing bio-informatics research. Unlike classification, clustering doesn't assign predefined categories; each cluster possesses distinct commonalities. The primary objective of Clustering techniques is to categorize data with Similar properties into groups, paving the way for further analysis using alternative methods. In recent years, propelled by Artificial Intelligence (AI) growth, clustering algorithms have gained prominence in machine learning. Scholars, both domestically and internationally, have delved deeper into research to enhance the processing capacity of large-scale data through innovative clustering algorithms. Notable advancements include granularity-based clustering algorithms, uncertainty-based clustering algorithms, entropy-based clustering algorithms, and integrated clustering algorithms. Beyond these mentioned approaches, many clustering algorithms exist, each with unique characteristics. It is crucial to recognize that no single algorithm is universally applicable; understanding the strengths and weaknesses of each algorithm is key to utilizing and researching them effectively.

3.8 CHALLENGES AND POTENTIAL SOLUTIONS

Machine learning is fundamental to data mining and is a widely employed method for Data computation. A notable potential of machine learning algorithms lies in their capacity to sift through extensive data, revealing patterns that might otherwise be overlooked. In biomedical research, particularly in the era of big data, Machine Learning (ML) plays a crucial role in uncovering Recognized patterns within biological systems. Yet, the current implementation of machine learning in biomedical data encounters multiple hurdles and difficulties.

3.9 UTILIZING MACHINE LEARNING FOR EXTRACTING INSIGHTS FROM DNA SEQUENCE DATA

Machine learning, a pivotal domain in computer science, serves a dual role. Firstly, it enables the extraction of valuable knowledge from vast datasets, facilitating data-driven insights. Simultaneously, various domains seek to derive actionable knowledge from data to inform their practices, presenting opportunities and challenges for machine learning's advancement [15]. The collaborative integration of machine learning with diverse fields has fostered a symbiotic relationship, significantly contributing to the growth and prosperity of machine learning deoxyribonucleic acid, a fundamental biological macromolecule and the building block of Genome, plays a crucial role in storing genetic information. With the ongoing development of genome sequencing systems, the focus in DNA sequence studies has shifted from accumulating raw data to interpreting it. This discussion delineates four key machine-learning applications in deoxyribonucleic acid sequence data alignment, clustering, classification and pattern mining. Each application is analyzed within its biological context, elucidating its significance. Lastly, a comprehensive overview of recent research in machine learning within this domain is systematically presented.

3.9.1 ALIGNING DNA SEQUENCES FOR COMPARATIVE ANALYSIS

Sequence alignment involves comparing two or more sequences to assess the arrangement of bases. Its primary purpose is to align Sequences with unknown functions to those with well-known sequences. The outcomes of Genome alignment provide insights into the similarity between sequences and their biological characteristics. In bioinformatics, sequence alignment analysis stands out as a fundamental and critical aspect, facilitating the anticipation of the structure and functionality of biological sequences as shown in Figure 3.4 taken from `https://dna-surveillance.auckland.ac.nz/hros001/alignment.png`.

Through sequence alignment analysis, researchers can delve into the evolutionary aspects of DNA, considering possibilities such as genetic material recombination and mutation. While the evolutionary process of deoxyribonucleic acid cannot be fully recovered, studying evolution through sequence alignment analysis allows the exploration of homology between DNA sequences. Sequence comparison can be categorized into double-sequence comparison and multi-sequence comparison, with the latter being an extension that presents increased alignment complexity as the number of sequences rises. The field of biological sequence alignment has matured, yielding many alignment tools like CLUSTAL, TCOFFEE, and MUSCLE. For instance, in our study, three deoxyribonucleic acid sequences of identical length were compared using CLUSTAL software. The local visualization of the results revealed matched regions, depicted in red, indicating a 54.35% sequence similarity among the three sequences. This example illustrates a simplified comparison, whereas real-world sequence comparisons are often more intricate. Dual sequence alignment was a starting point for early biological sequence alignment studies. The Needleman-Wunsch algorithm [9], also well-known as the global sequence alignment method and optimization identical algorithm, is a widely used sequence alignment technique that Needleman and Wunsch developed. It employed dynamic programming algorithms

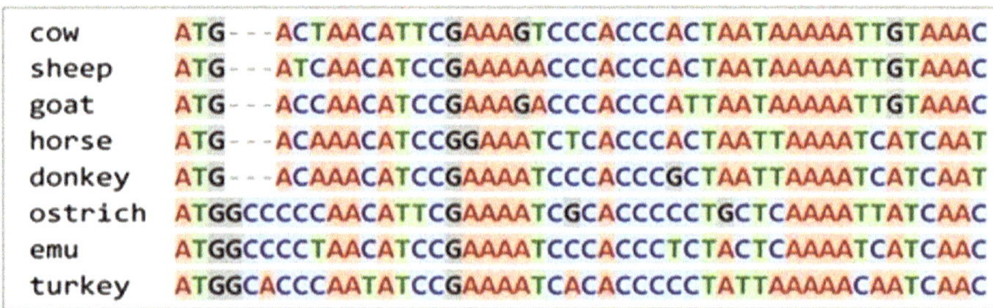

FIGURE 3.4 DNA sequence alignment of different species.

to achieve dual sequence alignment based on the similarity of the entire sequence. Smith and Waterman refined the dynamic programming approach to become a local optimum algorithm that can find sequence fragments that have a high degree of local similarity between two sequences. The Smith-Waterman algorithm's slow comparison speed is one of its drawbacks. Finding the longest common substring between two DNA sequences is necessary to find the maximum matching base number. First, compute the double sequence alignment score matrix to find the matched string. Next, apply the dynamic programming approach. We employed the CLUSTAL program to compare the two DNA sequences we had chosen to be of unequal length. The figure displays the local view of the comparison results. Blanks must be inserted to find the greatest number of matching bases because the two DNA sequences have different numbers of bases. The number of bases for a perfect match is Twenty-five, and the local similarity is also very high.

Later, BLAST and FASTA found crucial applications in querying and searching Genetic data repositories. As research deepened, swarm intelligence algorithms and their enhancements gradually found applications in aligning biological sequences, including genetic algorithms and ant colony algorithms [16]. Jangam et al. [17] proposed a hybrid algorithm for double sequence alignment, combining genetic and ant colony algorithms. This algorithm optimizes sequence alignment results by leveraging the ant colony algorithm's local attribute search potential and the Genome algorithm's global feature search capability. While effective for short and medium sequences, it faces challenges with slow convergence and low search efficiency for longer sequences.

To tackle the challenges of slow convergence and local optimization in the ant colony algorithm, Zhao et al. [18] proposed a novel approach by incorporating a sequence comparison method into an enhanced ant colony algorithm. This method involves adjusting ant positions and modifying pheromones at various intervals; the algorithm mitigates the risk of falling into local optima. However, the computational workload is substantial, leading to prolonged solution times. Multi-sequence alignment, an extension of double-sequence alignment, poses challenges related to excessive data storage and high computational complexity, particularly with many sequences. Given MSA's NP-complete nature, approximate alignment heuristic algorithms, such as CLUSTAL, T-Coffee, and CLUSTALW, have been developed. Iterative algorithms have gained prominence, with [19] proposing a graph-based DNA multi-sequence alignment algorithm called MWPAlign. Despite its linear time complexity, this algorithm struggles with varying sequence mutation rates.

Lee et al. [20] introduced a multi-sequence alignment genetic algorithm (GA-ACO) incorporating ant colony optimization. The hybrid approach combines genetic algorithms and ant colony optimization, addressing large-scale computation challenges but sacrificing search speed. Naznin et al. [21] presented a genetic algorithm vertical decomposition (VDGA) for multi-sequence alignment. VDGA effectively addresses the complexity of multiple sequence alignment by dividing sequences vertically, employing a guide tree method, and combining subsequences for a comprehensive alignment.

In the realm of heuristic techniques, stochastic methods like genetic algorithms (GA) have proven effective for solving MSA problems. Chowdhury and Garai [22] highlighted the advantages of genetic algorithms in MSA, emphasizing their ability to find optimal or suboptimal solutions regardless of sequence length and quantity. The genetic algorithm's versatility and potential for optimizing the objective function make it a valuable tool for sequence alignment.

As biological Genome sequence data grows, traditional sequence comparison tools face challenges in handling large datasets. Distributed computing frameworks like Hadoop have become essential for processing and storing big data. Mondal and Khatua [23] introduced a distributed sequence comparison algorithm, MRaligner, implemented in the Apache Spark framework using MapReduce. This algorithm significantly improves sequence comparison efficiency and demonstrates the potential for speeding up processing with scalable processors and effective distributed Hadoop Distributed File System (HDFS) management. Evaluating biological sequence alignment algorithms involves assessing efficiency and sensitivity to achieve optimal alignment results. While the Smith-Waterman algorithm for double-sequence alignment is extremely sensitive, its complexity is a drawback. Algorithms like FASTA and BLAST sacrifice some sensitivity for speed. CLUSTALW stands out among multiple sequence alignment algorithms for its everyday use and effectiveness. Challenges in sequence alignment revolve around improving sensitivity for sequences with substantial differences while maintaining algorithm efficiency. With the advent of next-generation sequencing technology (NGS), the field has witnessed a surge in biological data. Sequence alignment remains an indispensable step in deciphering relationships between sequences. However, traditional sequence alignment tools face inefficiencies in computing time for large input sequences. Future research in this field focuses on distributed sequence alignment and parallel computing to cope with high-throughput biological sequence data.

3.10 CATEGORIZING DNA SEQUENCES

In machine learning, classification is a crucial mining task that aims to create a model that predicts the category of unknown samples using a training sample set. Due to the non-numerical attributes of sequence elements, their relationships, and the various lengths of various sequences, the classification of biological sequences emerges as a challenging problem in data mining.

Different approaches have been proposed for DNA succession order. Directed Acyclic Word Graphs (DAWGs) were developed by Levy et al. [24] to classify DNA sequences. Ranawana et al. [25] proposed a multi-classifier system employing various coding techniques and neural networks, while Muller et al. [26] proposed a vector space approach for classification. Ma et al. [27] presented a method for classifying DNA sequences using a neural network and the expectation-maximization algorithm. This methodology, applied to distinguish E. coli advertisers, showed further developed execution on different datasets by avoiding suspicions about uniformly conveyed spacer lengths. Zaki et al. [28] presented a variable-request stowed away Markov model (VOGUE) for DNA succession grouping, using variable arrangement mining to extricate continuous examples with various lengths and spacings. VOGUE exhibited higher characterization precision than conventional wells. Recently, convolutional neural networks (CNNs), a well-known profound learning model, have been applied to DNA succession grouping. Nguyen et al. [29] proposed a strategy utilizing CNNs to group DNA successions as text information, accomplishing massive enhancements across various datasets. Regulated learning grouping undertakings in genomics frequently depend on highlight extraction. Bosco et al. [30] presented two profound learning models, showing their adequacy in extricating helpful highlights from input designs for order errands on different datasets. Because sequences lack clear features and conventional representation methods can result in high-dimensional problems, the feature selection process remains a major obstacle in genomics despite the advancements that have been made. Addressing the arrangement includes examining high-layered information to stay central focuses around the area of exploration.

3.11 DNA SEQUENCE CLUSTERING

In machine learning, classification is a crucial mining task that aims to create a model that predicts the category of unknown samples using a training sample set. Due to the non-numerical attributes of sequence elements, their relationships, and the various lengths of various sequences, the classification of biological sequences emerges as a challenging problem in data mining.

Different approaches have been proposed for DNA succession order. Directed Acyclic Word Graphs (DAWGs) were developed by Levy et al. [24] to classify DNA sequences. Ranawana et al. [25] proposed a multi-classifier system employing various coding techniques and neural networks, while Muller et al. [26] proposed a vector space approach for classification. Ma et al. [27] presented a method for classifying DNA sequences using a neural network and the expectation-maximization algorithm. This methodology, applied to distinguish E. coli advertisers, showed further developed execution on different datasets by avoiding suspicions about uniformly conveyed spacer lengths. Zaki et al. [28] presented a variable-request stowed away Markov model (VOGUE) for DNA succession grouping, using variable arrangement mining to extricate continuous examples with various lengths and spacings. VOGUE exhibited higher characterization precision than conventional wells. Recently, convolutional neural networks (CNNs), a well-known profound learning model, have been applied to DNA succession grouping. Nguyen et al. [29] proposed a strategy utilizing CNNs to group DNA successions as text information, accomplishing massive enhancements across various datasets. Regulated learning grouping undertakings in genomics frequently depend on highlight extraction. Bosco et al. [30] presented two profound learning models, showing their adequacy in extricating helpful highlights from input designs for order errands on different datasets. Because sequences lack clear features and conventional representation methods can result in high-dimensional problems, the feature selection process remains a major obstacle in genomics despite the advancements that have been made. Addressing the arrangement includes examining high-layered information to stay central focused around the area of exploration.

3.12 DNA SEQUENCE PATTERN MINING

Throughout DNA advancement, the protection of grouping designs is significant in the natural examination. These examples, frequently unambiguous practical pieces inside DNA groupings, assume a vital part in grasping the construction and capability of qualities. The recognizable proof of monitored districts and their ensuing development into particular succession designs is vital to DNA grouping information investigation. This cycle supports foreseeing the capability of DNA succession and clarifying the transformative connections between them. It is necessary to mine DNA sequence patterns to locate genes and comprehend their functions.

The idea of revamped grouping design mining, presented by Srikant et al. [31], has since developed into a crucial region inside AI, earning impressive consideration from scientists. Different kinds of substitution designs, for example, exchange thing sets, tedious after-effects, and substitution foundations add to the investigation of DNA succession designs. A schematic outline represents the grouping model in Figure 3.5 [32]. The eight green lines address unmistakable groupings, while the three diversely shaded squares mean three particular succession designs. Pattern one can be found in sequences 3, 5, 6, and 7, pattern two can be found in sequences 1, 2, 4, 5, 6, and 8, and pattern three can only be found in sequence 3. Further investigation reveals that sequences 4 and 8 contain two patterns, necessitating a more in-depth look at their shared characteristics. The Apriori calculation is a device generally involved in mining information affiliation rules, revealing oftentimes happening informational collections to help direction. The Generalized Sequential Patterns (GSP) algorithm, an Apriori-based approach that introduces time and conceptual constraints, further developed this idea. Regardless of its viability, the GSP calculation faces proficiency challenges while managing enormous scope arrangement information bases because of the age of various applicant designs, requiring successive data set checks. As a result, researchers frequently combine Apriori principles with other algorithms to increase overall efficacy and produce more reliable

FIGURE 3.5 DNA sequence patterns.

results. Two main types of calculations can be used to investigate patterns in biological sequence. Heuristic hunt techniques, including iterative cycles, offer diminished computational intricacy yet may take a chance with neighbourhood improvement. Alternately, thorough inquiry procedures consider every conceivable arrangement, assessing them separately to distinguish the ideal arrangement. The decision between these techniques relies upon the particular prerequisites of the DNA arrangement investigation, with heuristic systems leaning toward partitioned successions and comprehensive methodologies pertinent when a thorough assessment of all prospects is fundamental.

Two main types of sequential pattern mining algorithms are currently in use. Some emphasize mining elective examples inside individual groupings, surpassing a predetermined edge. However, a different approach is required because biological sequence analysis frequently requires the simultaneous examination of patterns across sequence sets. Calculations in the subsequent class have practical experience in mining rehashed succession designs in informational indexes, yet their effectiveness reduces when stood up to with gigantic organic information. A few specialists have proposed creative calculations to address these difficulties. Zhou et al. [33] presented mMBioPM, upgrading hash table designs and diminishing overt repetitiveness in mining results. Chen and Liu's FBPM algorithm utilized a pattern growth technique and the main mode concept for effective mining of biological sequence frequent patterns. The SPMM algorithm, developed by [14] to speed up and improve the quality of sequence pattern mining, incorporates a Markov chain strategy to overcome issues with memory and time. Guojun Mao [34] presented an affiliation grid, a conservative information structure working with successfully mining key parts in DNA groupings. This clever construction handles super lengthy arrangements with restricted extra room, giving a productive method for dissecting DNA groupings. Customary DNA succession design mining calculations frequently yield repetitive, organically irrelevant examples. Generating excessive candidate patterns requires a significant amount of time and space, and these brief patterns contribute to inefficiency. Planning further developed search methodologies and dispensing with repetitive examples will be critical for propelling DNA grouping design mining research from here on out.

3.13 CURRENT CHALLENGES

DNA grouping examination offers a significant road for investigating hereditary varieties among creatures. Advanced analysis techniques are in high demand due to the exponential growth of DNA sequence data. However, DNA sequence data mining still faces several obstacles:

1. Productivity obstacles continue dealing with huge scope DNA succession information, representing a test to ideal and powerful handling.
2. Fitting DNA grouping information mining calculations to explicit organic necessities requires a nuanced approach, considering the different foundation information and qualities of successions.

3. Extricating significant grouping qualities from DNA arrangements and contriving powerful comparability measures for surveying succession likeness stay basic parts of the examination.
4. The inborn "discovery" nature of AI represents a test in giving naturally significant clarifications to the model results, restricting the more extensive utilization of these models in specific settings

3.14 CONCLUSION

Data collection in various life science applications, such as omics, biological imaging, and medical imaging, has been transformed by rapid hardware technology advancements in recent decades. However, advancements in life science technology have also posed significant obstacles. At present, a point of convergence of examination lies in the use of various information mining advances to bio-informatics examination, enveloping the improvement of information mining designs, AI calculations, and novel information-digging capabilities custom-made for natural data handling. The interdisciplinary methodology has prodded the development of AI, seeing forward leaps in fake brain organizations, profound learning, and support realizing, all worked with by expanded registering power, speeding up information capacity speeds, and diminished figuring costs. Scientists from various fields are utilizing these technologies to analyze biological data, anticipating mining outcomes that are more meaningful and beneficial to societal advancement.

Based on this body of research, we've identified two essential aspects of the application of machine learning to DNA sequence analysis that require special attention: First and foremost, there is a need to explain the natural meaning of DNA successions. Even though many algorithms work well when analyzing DNA sequences, their results are often very sensitive and specific, making it hard to use them in real-world situations. Tending to this challenge includes incorporating the natural setting of DNA arrangements into the information mining process, justifying centred consideration and examination. Also, as information volumes continue to increase, conventional examination apparatuses face failures in figuring time. Research endeavours should focus on planning effective estimation techniques, with the combination of dispersed registration and equal processing promising significant enhancements in mining proficiency. It is crucial to choose the right DNA sequence coding method for a given task simultaneously. This choice improves algorithm performance and cuts down on training time.

In conclusion, aspects like sequencing technology, data structure, and sequence similarity are covered in this comprehensive review of the DNA sequence data source and characteristics. We give a short outline of AI calculations, featuring difficulties these calculations face in natural succession information mining and proposing possible future arrangements. The review delves into alignment, classification, clustering, and pattern mining—four typical applications of machine learning in DNA sequence data—to examine their biological significance and background. We provide a methodical summary of recent research by scholars from the United States and other countries, highlighting important issues and offering suggestions for future research. Expecting a more consistent combination of science and AI, we imagine the fulfilment of more powerful mining brings about future examination.

REFERENCES

1. Ranjeet Kumar Rout, Saiyed Umer, Monika Khandelwal, Smitarani Pati, Saurav Mallik, Bunil Kumar Balabantaray, and Hong Qin. Identification of discriminant features from stationary pattern of nucleotide bases and their application to essential gene classification. *Frontiers in Genetics*, 14:1154120, 2023.
2. Monika Khandelwal, Sabha Sheikh, Ranjeet Kumar Rout, Saiyed Umer, Saurav Mallik, and Zhongming Zhao. Unsupervised learning for feature representation using spatial distribution of amino acids in aldehyde dehydrogenase (aldh2) protein sequences. *Mathematics*, 10(13):2228, 2022.
3. Sang Tae Park and Jayoung Kim. Trends in next-generation sequencing and a new era for whole genome sequencing. *International neurourology journal*, 20(Suppl 2):S76, 2016.
4. Fred Sanger and Alan R Coulson. A rapid method for determining sequences in DNA by primed synthesis with DNA polymerase. *Journal of molecular biology*, 94(3):441–448, 1975.
5. Allan M Maxam and Walter Gilbert. A new method for sequencing DNA. *Proceedings of the National Academy of Sciences*, 74(2):560–564, 1977.
6. William R Pearson. An introduction to sequence similarity ("homology") searching. *Current protocols in bioinformatics*, 42(1): Chapter 3.1.1–3.1.8, 2013.
7. Igor B Rogozin, Luciano Milanesi, and Nikolay A Kolchanov. Gene structure prediction using information on homologous protein sequence. *Bioinformatics*, 12(3):161–170, 1996.
8. Steven Henikoff and Jorja G Henikoff. Amino acid substitution matrices from protein blocks. *Proceedings of the National Academy of Sciences*, 89(22):10915–10919, 1992.
9. Amr Ezz El-Din Rashed, Hanan M Amer, Mervat El-Seddek, and Hossam El-Din Moustafa. Sequence alignment using machine learning-based needleman–wunsch algorithm. *IEEE Access*, 9:109522–109535, 2021.
10. William R Pearson and David J Lipman. Improved tools for biological sequence comparison. *Proceedings of the National Academy of Sciences*, 85(8):2444–2448, 1988.
11. Emre Delibaş and Ahmet Arslan. Dna sequence similarity analysis using image texture analysis based on first-order statistics. *Journal of Molecular Graphics and Modelling*, 99:107603, 2020.
12. Ranjeet Kumar Rout, Saiyed Umer, Sabha Sheikh, and Amrit Lal Sangal. *Artificial Intelligence Technologies for Computational Biology*. CRC Press, 2022.
13. Monika Khandelwal and Ranjeet Kumar Rout. Prmxai: protein arginine methylation sites prediction based on amino acid spatial distribution using explainable artificial intelligence. *BMC bioinformatics*, 24(1):376, 2023.
14. Zhang Junyan and Yang Chenhui. Sequence pattern mining based on markov chain. In *2015 7th International Conference on Information Technology in Medicine and Education (ITME)*, pages 234–238. IEEE, 2015.
15. Monika Khandelwal, Ranjeet Kumar Rout, and Saiyed Umer. Prediction of protein-protein interaction using support vector machine based on spatial distribution of amino

acids. In *International Conference on Advances and Applications of Artificial Intelligence and Machine Learning*, pages 23–32. Springer, 2022.

16. Biswa Mohan Sahoo, Ranjeet Kumar Rout, Saiyed Umer, and Hari Mohan Pandey. Ant colony optimization based optimal path selection and data gathering in wsn. In *2020 International Conference on Computation, Automation and Knowledge Management (ICCAKM)*, pages 113–119. IEEE, 2020.
17. Sujit R Jangam and Nirupam Chakraborti. A novel method for alignment of two nucleic acid sequences using ant colony optimization and genetic algorithms. *Applied Soft Computing*, 7(3):1121–1130, 2007.
18. Yidan Zhao, Ping Ma, Jie Lan, Chun Liang, and Guoli Ji. An improved ant colony algorithm for DNA sequence alignment. In *2008 international symposium on information science and engineering*, volume 2, pages 683–688. IEEE, 2008.
19. Z Xiao. A multiple alignment approach for DNA sequence based on the maximum weighed path algorithms. *Xi'an University of Electronic Science and Technology*, pages 8–9, 2006.
20. Zne-Jung Lee, Shun-Feng Su, Chen-Chia Chuang, and Kuan-Hung Liu. Genetic algorithm with ant colony optimization (ga-aco) for multiple sequence alignment. *Applied Soft Computing*, 8(1):55–78, 2008.
21. Farhana Naznin, Ruhul Sarker, and Daryl Essam. Vertical decomposition with genetic algorithm for multiple sequence alignment. *BMC bioinformatics*, 12:1–26, 2011.
22. Biswanath Chowdhury and Gautam Garai. A review on multiple sequence alignment from the perspective of genetic algorithm. *Genomics*, 109(5-6):419–431, 2017.
23. Sudip Mondal and Sunirmal Khatua. Accelerating pairwise sequence alignment algorithm by mapreduce technique for next-generation sequencing (ngs) data analysis. In *Emerging Technologies in Data Mining and Information Security: Proceedings of IEMIS 2018, Volume 2*, pages 213–220. Springer, 2019.
24. Samuel Levy and Gary D Stormo. Dna sequence classification using dawgs. *Structures in Logic and Computer Science: A Selection of Essays in Honor of A. Ehrenfeucht*, pages 339–352, 2005.
25. Romesh Ranawana and Vasile Palade. A neural network based multi-classifier system for gene identification in DNA sequences. *Neural Computing & Applications*, 14:122–131, 2005.
26. H-M Muller and SE Koonin. Vector space classification of DNA sequences. *Journal of theoretical biology*, 223(2):161–169, 2003.
27. Qicheng Ma, Jason TL Wang, Dennis Shasha, and Cathy H Wu. Dna sequence classification via an expectation maximization algorithm and neural networks: a case study. *IEEE Transactions on Systems, Man, and Cybernetics, Part C (Applications and Reviews)*, 31(4):468–475, 2001.
28. Mohammed J Zaki, Christopher D Carothers, and Boleslaw K Szymanski. Vogue: A variable order hidden markov model with duration based on frequent sequence mining. *ACM Transactions on Knowledge Discovery from Data (TKDD)*, 4(1):1–31, 2010.
29. Ngoc G Nguyen, Vu Anh Tran, Dau Phan, Favorisen R Lumbanraja, Mohammad Reza Faisal, Bahriddin Abapihi, Mamoru Kubo, and Kenji Satou. Dna sequence classification by convolutional neural network. *Journal Biomedical Science and Engineering*, 9(5):280–286, 2016.
30. Giosuè Lo Bosco and Mattia Antonino Di Gangi. Deep learning architectures for DNA sequence classification. In *Fuzzy Logic and Soft Computing Applications: 11th International Workshop, WILF 2016, Naples, Italy, December 19–21, 2016, Revised Selected Papers 11*, pages 162–171. Springer, 2017.

31. Ramakrishnan Srikant and Rakesh Agrawal. Mining sequential patterns: Generalizations and performance improvements. In *International conference on extending database technology*, pages 1–17. Springer, 1996.
32. Aimin Yang, Wei Zhang, Jiahao Wang, Ke Yang, Yang Han, and Limin Zhang. Review on the application of machine learning algorithms in the sequence data mining of DNA. *Frontiers in Bioengineering and Biotechnology*, 8:1032, 2020.
33. Qingda Zhou, Qingshan Jiang, Sheng Li, Xiaobiao Xie, and Lida Lin. An efficient algorithm for protein sequence pattern mining. In *2010 5th International Conference on Computer Science & Education*, pages 1876–1881. IEEE, 2010.
34. Guojun Mao. Association matrix method and its applications in mining DNA sequences. In *Advances in Artificial Intelligence, Software and Systems Engineering: Proceedings of the AHFE 2019 International Conference on Human Factors in Artificial Intelligence and Social Computing, July 24-28, 2019, Washington DC, USA 10*, pages 154–159. Springer, 2020.

4 Efficient Detection and Recuperation of Mental Health using X (Formerly Twitter) and Fitbit Data-Based Recommendation System

Shikha Mehta, Parul Agarwal, Kunwar Vidhan, Praveen Pathak, Amogh Misra, and Savvy Bansal

4.1 INTRODUCTION

According to the World Health Organization (WHO), 75% of people with a neurological disorder in developing countries do not receive proper treatment [1]. This is due to the limited, affordable medical facilities available. Nearly 1 billion people are affected by mental disorders. Due to the rise of the COVID-19 pandemic around the world, the mental health of people is further impacted and has become an ever-increasing problem in our world nowadays. Mental health is related to the brain, which controls various human activities. To attain resilience, it is necessary to balance the various functions and activities. Mental health could affect the daily life of many people, including their relationships, living conditions, and physical health. Human beings can live a happy and healthy life. Different mental problems like depression, trauma, and tension can affect day-to-day life [2]. Thus, its preservation is important for every human being.

Mental health [3] is a contemporary issue, and people are now focusing more on achieving their mental peace instead of overdoing or overthinking others or something else. But in this process, they sometimes have to get through rough patches, or it can be said, long periods where they are extremely low on energy. This could be because of stress, anxiety, or any other reason. There should be a platform that could detect a person's mood and then recommend to the user some ways to lighten their mood and cheer them up. The applications existing in this field are not very

DOI: 10.1201/9781032660714-4

convenient as they want the user to enter their data to predict the mood, which becomes monotonous for the user at their low points. There should be an application that automates the prediction process and then helps in recovery. There is an urgent need for applications that can detect the user's mood and then help them in the recovery process. It is important because people tend to take drastic measures in an emergency or a dire situation. These measures can often cause problems for them or their loved ones. This application could be handy to reduce these risks. Whenever the user is in a bad or sad mood, they can just sit and use the application for recommendations based on their mood. This application could help the user to go into their happy place and would help them to recover from their low points. This increases the efficacy of the users in their day-to-day work and would also not put anyone's life on the line by preventing the users from taking any drastic actions. Another factor that would entertain is the availability of such applications that could be used for practical purposes. The existing applications are mainly rule-based systems, so the user has to answer a set of questions every time. That is not very helpful because the user often feels reluctant to answer any questions when they are feeling low. Also, this can infuriate the cause and develop a greater tendency to attempt drastic measures.

Several researchers are trying to explore technology-driven solutions for mental health detection. Cognitive Behavioural Therapy (CBT) [4] is the popular technique used to treat patients with depressive symptoms. According to studies, the Behavioural Activation (BA) component of CBT provides the same positive effects as full CBT [5] and is simpler as compared to CBT [6]. BA involves engaging the individual in or activating in more pleasant activities. This implies changing the mood or distracting the individual from their current mental state. In [7], authors developed mobile applications to identify the mood and recommended routine activities as an intervention. In [8], authors developed the HEADGEAR app for controlling depression symptoms. In [9], authors assessed the usability of the mobile app "Bluewatch" developed as a self-guided Health intervention. Various other apps [10–12] have been developed to understand the mental health of patients through questionnaires and then provide recommendations for behavioral activations. However, all these applications are based on the assumption that an individual is aware of their mental health being not well. Further, these studies have been done in the foreign context only. Another assumption is that an individual is ready to complete the survey or questionnaire to reveal the personalized interventions that suit the individual most. These questions sometimes push the user to close the application and do some other work. On the contrary, in the real world, neither the individual nor the acquaintances are aware of the mental illness of the individual, which sometimes leads to severe effects. To handle this situation, there is a need to devise a mechanism for early detection of mental health conditions for timely interventions.

Researchers have performed various works on mood prediction and recommendation systems. Mikus et al. [13] uses a recurrent neural network to predict the short-term moods of the patients. The authors have collected data from various patients diagnosed with depression. However, the authors collected data traditionally and did not provide any recovery procedure. Another work by Cernianet al. [14] identifies the mood of a person based on external parameters like pulse, skin electroconductivity (similar to galvanic skin rate), and temperature. The paper works on a machine learning algorithm, such as Support Vector Machine (SVM), to identify the mood of users. However, some other classification algorithms could be used to get better prediction [15–17]. A work on pattern classification is performed by [18]. This paper deals with functions like the polynomial kernel function, the sigmoid kernel function, and the radial basis function kernel, which produce non-linear decision boundaries. It is used in mood classification. The training data in the dataset are sampled based on an unknown probability distribution and a generic loss function used for calculating error. A novel feature extraction method is presented for music classification based on mood [19]. This paper discusses a novel technique in the field of music mood classification, with a relationship to emotions, which is very useful for music retrieval or recommendation.

Clustering of audio attributes is helpful in the generation of a playlist based on a person's mood [20]. In this approach, clusters are first created by extracting the data from the user. This data is the music that the user likes to listen. Each cluster will contain similar feature audio tracks. These clusters represent the mood of users. Since each cluster contains a playlist based on the mood, playing music one after the other will enhance user satisfaction. This approach assigns the cluster to the unknown user based on present emotions and recommends an appropriate track. A web-based API for specifying the data analysis is given by Sciandraand Spera [21]. Spotify (Spotify Web API, 2019) is a well-known app in the world. The authors proposed a model for handling the popularity index of the song. Zenonos et al. [22] introduced a mood detection model in a work atmosphere.

Various studies have been performed on recommendation systems. [23] aimed to discover the impact of personality traits in user-to-user filtering, i.e., collaborative filtering. Here, the categorization of personality traits is done through the big five personality traits of psychology used in day-to-day life. The data is pre-processed and trained using the Naive Bayes classifier. In the testing phase, music is recommended, and feedback is again sent to the database. The input data is taken from the Facebook status of the people for the classification process.

Some recommendation works are given for groups. Agarwal et al. [24] proposed a recommendation algorithm using Ant colony optimization for groups. Another work on evaluating group recommendation is given in [25]. The method used in this work records the individual and shared preferences of individuals in groups and compares the performance of both types of recommendation systems over a wide range of music group recommendations.

Fuzzy models are also being employed by researchers in the recommendation system. Kasinatha et al. [26] proposed a system that has a smart music recommendation system using a fuzzy inference model. The system makes use of user's preferences, mood and time to recommend appropriate songs to them. Dong and Gu [27] use the music itself to make the model a more efficient and accurate recommendation system. The system is based on the fusion of the deep learning model. The model is validated through the ten-fold testing scheme. The class of prediction is pop, classic, and country music.

Another work on the music recommendation system [28] deals with new and innovative ideas to improve the current music recommender system. The paper deals with solving some of the challenges that occur in recommending the music system. Some challenges are cold start problems, automatic playlist continuation, evaluation of the recommendation system, etc. It also details the possible future direction and vision for the up-gradation of the upcoming music recommendations. There are some studies on heartbeat analysis [29]. The paper presents a study of the brain and heart's interaction between animals and humans. The authors use Detrended fluctuation Analysis (DFA). The paper illustrates that DFA technology could be a useful biomedical method for measuring the heart's state and thus shows a direct relation between the heart rate and the psychological state of the human and animal mind. The results of the paper show that stress and fear behaviour is directly related to the biogenic monoamine nervous system. Ronaldo and Silaghi [30] discuss the initial methodology used by YouTube for a recommendation. The authors improved the initial method, which not only improved the recommendation system but also overcame the anomaly of video recommendation and false comment creation over videos. Also, it aided YouTube in quantifying the meaning of a video's recommendation to an individual. Thus, a variety of work has been performed, but none of the work focuses on both detection and recovery combined. Also, most of the works either take up existing data or collect data from the user by questioning.

These days, most individuals wear health bands to keep track of their health for timely interventions. Statistics show that in 2019, there were 29.57 million people who were actively using Fitbit products (`https://www.businessofapps.com/data/fitbit-statistics/`). The output of these products can also be tracked using associated web applications. Besides, social media websites like Facebook, Twitter (currently X), etc., have become popular platforms for expressing moods in the form of textual emotions and taking suggestions these days. Mood is the primary indicator of whether women are undergoing any mental stress, anxiety, etc. This model detects mood by taking physical parameters as input from a health band, mainly Fitbit, and the individual's social media posts. The mathematical model would be developed to devise the relationship between physical parameters, social media posts, and moods of the women. This mathematical model would be used to detect moods and understand the mental health of women. Subsequently, the recommender system would generate recommendations or Behavioural Activation (BA) to change the mood. The behavioural activations would include the playing of appropriate music on mobile as well as sending messages to acquaintances to make calls and take necessary steps. The overall objectives of the work are as follows:

1. Collect physical fitness from Fitbit data such as physical parameters like galvanic skin response (GSR), temperature, heart rate, calories, and social media data in the form of textual emotions.
2. Pre-processing of data to make it suitable for analysis and modeling.
3. To device the relationship between Fitbit parameters, social media posts, and mood of an individual.

4. Develop a hybrid recommender system to recommend personalized and tailored Behavioral Activations (BA) or engagements in the form of music to change the mood by relieving stress.

The work is divided into two phases-Mental Health Detection and Recommendation of behavioral activation [31] phase. In the first phase, data is being extracted in the form of textual emotions from social media platforms and physical parameter data from the details of the Fitbit device. Subsequently, data is preprocessed to make it suitable for developing the model. Initially, both the textual emotion data and Fitbit data are analyzed separately. For sentiment analysis of textual [32] data, TF-IDF and bag of words techniques are applied. For Fitbit data, a modified deep neural network classification model is proposed to improve the accuracy of the model. Subsequently, both the models are integrated to develop the hybrid recommendation system. Finally, in Recommendation of behavioral activation phase, a prediction model is applied to predict the mood of the user based on various parameters. Recommendations or behavioral activations are provided in the form of music or songs to change the mood of the users. The main challenge in this work was to collect Fitbit data of users as people are reluctant to share their details and observations of Fitbit data. Another challenge was to assess the accuracy of the proposed mental health recovery recommendation system. The effectiveness of the proposed model is evaluated by taking feedback from the users for the recommendations generated automatically. The chapter is divided into the following sections: Section 2 presents related work. Section 3 portrays the proposed hybrid model for mental health recuperation recommendation system followed by experiments and results in section 4. Conclusion and limitations are given in section 5.

4.2 PROPOSED METHODOLOGY

A Recommendation System [33], or a recommender system, is a subclass of Information Filtering System (IFS) whose aim is to predict the "preference" that a particular user gives to an item. They are most commonly used as playlist generators for music, videos and are used by big companies like Google (YouTube) and Netflix. They either make use of collaborative filtering or content-based filtering or can be a combination of both. The model is then used to predict items that the user has interest using the property of similarity between the available choices. One recommender system could be a combination of two or more previously made systems and are known as hybrid recommendation systems.

The detailed working of the proposed hybrid recommendation system is as follows: It begins with the collection of data which plays a critical role in the success of any recommender system. Most of the existing models focus on the social media data or ask several questions from the users for understanding their mental health. Asking numerous questions from a person to understand mental health does not seem to be a fair idea. Therefore, an automatic system as shown in Figure 4.1 is developed which understands or detects the mental health of the person from its routine activities like

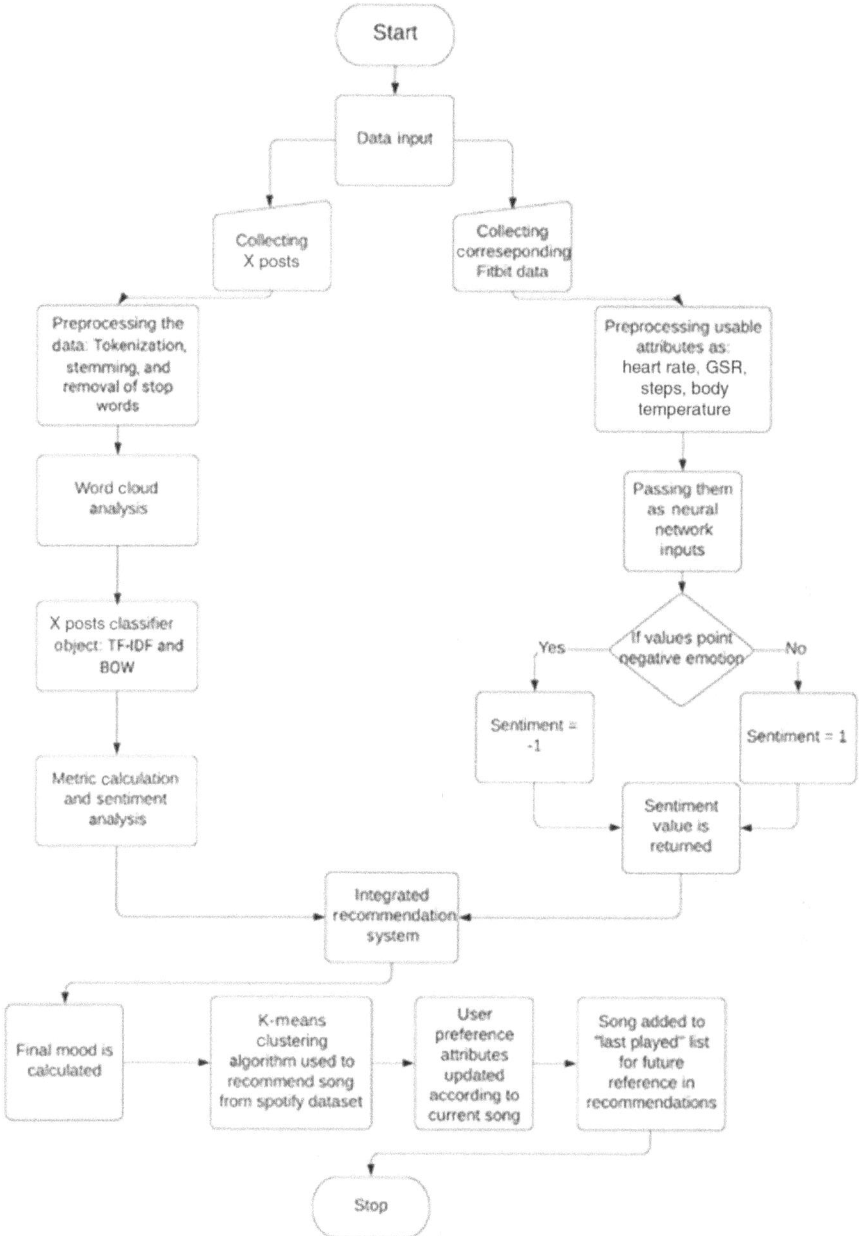

FIGURE 4.1 Architecture of mental health discovery and recuperation recommendation system.

his posts on social media and the physical activities and then recommends behavioral activations to change the mood of the user.

The complete working of the system is divided into two phases-Mental Health detection (MHD) and Recommendation of behavioral activation(RBA) phase. The first phase involves the development of the model for mental health discovery which involves understanding the data, preprocessing, and developing the mood classification model. This step is crucial to the system as the accuracy of recommendation completely depends on the accuracy of mental health detection. In the online phase, the system understands the mental state of the new users and generates behavioral activations or recommendations in the form of songs to change the mood of the users.

The various steps involved in the MHD phase is as follows:

4.2.1 DATA EXTRACTION

In this step, for the identification of textual emotions, Twitter (currently X) sentiment data was downloaded from the `https://www.kaggle.com/datasets/kazanova/sentiment140`. For information about physical parameters, Fitbit data available on the `https://www.kaggle.com/datasets/arashnic/fitbit` was downloaded. This data consisted of two files- one with information about physical parameters and the second file contained sentiments associated with the physical parameters. Both these files were combined to retain only the data with physical parameters and their associated emotions that are moods of the users. The various physical parameters used for developing the model include GSR, heart rate, the temperature of the body, steps, and calories.

1. **Steps:** It represents the steps moved by a person as per the watch data. This work is based on the assumption that a person is always walking with his Fitbit band.
2. **GSR:** It stands for galvanic skin response. The galvanic skin response (GSR), also known as Skin Conductance (SC), refers to changes in sweat gland activity, which reflects the intensity of participants' emotional state – or 'emotional arousal'.
3. **Calories:** Calories are the measure of the amount of energy consumed by a human body. The number of calories burnt is directly dependent on physical fitness as well as the mental fitness of the person.
4. **Temperature:** The average normal body temperature is generally accepted as 98.6°F (37°C). Some studies have shown that the "normal" body temperature can have a wide range, from 97°F (36.1°C) to 99°F (37.2°C). A temperature over 100.4°F (38°C) most often means that fever is caused by an infection or illness.

All these attributes are very relevant to mental health and thus they are used for the prediction of mood. The dataset with these parameters is used for training the model. The main challenge in this step was to prepare the data for testing the recommendation system. The testing of the proposed recommendation system required both types of data –tweets and physical parameters data from the Fitbit device of the same users. Therefore, the first step was to identify the users who use the Fitbit device and then pursue them to share their personal ID to access their Fitbit device and extract the data. After huge efforts, we were able to convince 100 subjects with Fitbit devices to share their personal Id in order to extract the data of physical parameters. To analyze textual emotions, tweets from the same set of users were downloaded from Twitter (currently X).

4.2.2 DATA PREPROCESSING

The proposed hybrid model uses two types of inputs-mental health expressed in terms of textual emotions captured from the tweets posted by the user and the physical parameters information extracted from the Fitbit device. Both these inputs involve different types of data that is Twitter (currently X) data is unstructured and data from the Fitbit device is in the structured form. Therefore, different types of pre-processing techniques are used for both types of data to make it suitable for developing the model.

4.2.2.1 X (Formerly Twitter) Data Pre-Processing

(a) Tokenization, Stemming, and Removal of stop words: These are the primary techniques for pre-processing the textual data. The process of tokenization is applied to segment the text into words, clauses, or sentences. This process also involves separating the words and removal of punctuations. Next stemming is performed which helps in reducing the related words to a common word or stem. Finally, the commonly used words also called stop words are removed as they are least useful for learning and may only increase the computational complexity of the task.

(b) Word Cloud Analysis: A Word Cloud is a visual representation of textual data. It helps in visualizing the importance of different words in the text based on their size and color shown in the cloud. The exponential growth of unstructured data in the form of text over social media platforms has escalated the need to analyze the huge amount of text. A Word Cloud is considered to be a popular choice these days as it provides a visual interpretation of the text and helps in gaining quick insights into the most prominent emotions/sentiments in a given text, by visualizing the word frequency in the text as a weighted list.

4.2.2.2 Fitbit Data Preprocessing

This dataset consists of a log of emotions and a log of physical attributes corresponding to it. In this dataset, many attributes had "NA" values that may have been recorded while the subject was asleep. Hence, the first step was to discard such extra entries that do not contribute to the prediction model. After eliminating the extra tuples, the data types of all the attributes were changed to their corresponding types. Specifically, the dates, originally stored as strings, were converted to the `DateTime` datatype. Similarly, attributes such as GSR (Galvanic Skin Response), Heart Rate, Calories, and Temperature were converted from `string` to `float`. Following this transformation, the data was normalized to scale all values to a common range, making the prediction model more unbiased. The goal of normalization is to standardize numeric columns without distorting differences in their ranges or losing information. The different types of scalers used are:

1. **Standard Scaler:** This scaler removes the mean and scales the data to unit variance. However, standard scaling does not guarantee balanced feature scales in the presence of outliers.
2. **Min-Max Scaler:** This technique scales the dataset so that all feature values are within the range [0,1]. It is a popular choice among researchers as it preserves the shape of the dataset without distortion. The transformation is computed using Eq. 4.1.

$$x' = \frac{x - \min(x)}{\max(x) - \min(x)} \tag{4.1}$$

   Here, x represents the original value of a parameter at a particular instant, $\min(x)$ and $\max(x)$ denote the minimum and maximum values in the dataset, respectively, and x' represents the normalized value.

4.2.3 DEVELOPMENT OF CLASSIFICATION MODEL

4.2.3.1 Model for Mood Prediction using Textual Emotions

The algorithm used to predict the mood or mental state of the user is Term Frequency-Inverse Document Frequency (TF-IDF). It is an information retrieval technique [34] that weights the term frequency (TF) of a word against its inverse document frequency (IDF). Here, a document refers to any text, such as a tweet on Twitter (currently X), a post on Facebook, or a news article. Each word or term in the text is associated with its respective TF and IDF scores. The product of the TF and IDF scores of a term in the analyzed text (tweets in this work) is called the TF-IDF weight, as shown in Eq. 4.2.

$$\text{tfidf}(t,d,D) = \text{tf}(t,d) \cdot \text{idf}(t,D) \tag{4.2}$$

Here, the term $\text{tf}(t,d)$ represents the frequency of the term, which is computed using Eq. 4.3.

$$\text{tf}(t,d) = \frac{f_{t,d}}{\sum_{t' \in d} f_{t',d}} \tag{4.3}$$

where $f_{t,d}$ refers to the count of a term in the text or document, i.e., the number of times a term t appears in document d. On the other hand, inverse document frequency (IDF) measures the importance of a term by determining whether it is rare or frequent across documents. It is computed using Eq. 4.4, where the numerator represents the total number of documents, and the denominator represents the number of documents containing the word. A higher TF-IDF score (weight) indicates that the term is rarer in a given document.

$$\text{idf}(t,D) = \log\left(\frac{N}{|\{d \in D : t \in d\}|}\right) \tag{4.4}$$

The detailed steps for developing the classification model using the TF-IDF algorithm [34] are as follows:

1. Compute the term frequencies of positive and depressive tweets, denoted as $\text{tf}_{\text{positive}}$ and $\text{tf}_{\text{negative}}$, respectively. Compute their inverse document frequencies as $\text{idf}_{\text{positive}}$ and $\text{idf}_{\text{negative}}$.
2. Compute the probability of each word in the processed text using:

$$\text{prob}_{\text{depressive}}(\text{word}) = \text{tf}_{\text{depressive}}(\text{word}) \cdot \log\left(\frac{\text{total˙count}_{\text{depressive}} + \text{total˙count}_{\text{positive}}}{\text{idf}_{\text{depressive}}(\text{word}) + \text{idf}_{\text{positive}}(\text{word})}\right) \tag{4.5}$$

$$\text{prob}_{\text{positive}}(\text{word}) = \text{tf}_{\text{positive}}(\text{word}) \cdot \log\left(\frac{\text{total˙count}_{\text{depressive}} + \text{total˙count}_{\text{positive}}}{\text{idf}_{\text{depressive}}(\text{word}) + \text{idf}_{\text{positive}}(\text{word})}\right) \tag{4.6}$$

   Compute the total probability of depressive words as $\text{sum˙probable}_{\text{depressive}}$ and the total probability of positive words as $\text{sum˙probable}_{\text{positive}}$.
3. Compute the average probability for each word:

$$\text{Avg˙prob}_{\text{depressive}}(\text{word}) = \frac{\text{prob}_{\text{depressive}}(\text{word}) + 1}{\text{sum˙probable}_{\text{depressive}} + \text{total˙depressive˙words}} \tag{4.7}$$

$$\text{Avg˙prob}_{\text{positive}}(\text{word}) = \frac{\text{prob}_{\text{positive}}(\text{word}) + 1}{\text{sum˙probable}_{\text{positive}} + \text{total˙positive˙words}} \tag{4.8}$$

Here, $sum\_probable_{depressive}$ represents the sum of all probable depressive word values, and $total\_depressive\_words$ is the number of depressive words. Similarly, $sum\_probable_{positive}$ represents the sum of all probable positive word values, and $total\_positive\_words$ is the number of positive words.

4. Perform classification using the following algorithm:

pdepressive = 0
ppositive = 0
for each word in text **do**
  if word is present in $prob_{depressive}$ **then**
    $pdepressive += \log(prob_{depressive}(word))$
  else
    $pdepressive -= \log(prob_{depressive}(word))$
  end if
  if word is present in $prob_{positive}$ **then**
    $ppositive += \log(prob_{positive}(word))$
  else
    $ppositive -= \log(prob_{positive}(word))$
  end if
  if pdepressive > ppositive **then**
    Text is classified as **depressive**
  else
    Text is classified as **positive**
  end if
end for

4.2.3.2 Modified Neural Network algorithm for Mood Prediction using Physical Parameters

A neural network is a series of algorithms that endeavors to recognize underlying relationships in a set of data through a process that mimics the way the human brain operates. In this sense, neural networks refer to systems of neurons, either organic or artificial. In this work, a modified sequential neural network is developed for improving the accuracy of mood classification using physical parameters. The architecture of the proposed modified neural network algorithm is as follows:

1. **Dense Layer:** The layer consists of 5 neurons, equal to the number of the attributes in the input, since the Tanh function maps the negative value to strongly negative and similarly positive to strongly positive value the initial results of the layers are enhanced. Here, the focus is on the zero-centered outputs produced by this function.
2. **Dropout Layer:** This layer is added to increase the probability of switching the layers off and in turn producing results that are less biased and are less likely to be overfitted.
3. **Dense layer:** This layer consists of 10 nodes, and has the activation function as Tanh.

4. **Dense layer:** The layer consists of 7 neurons, the Tanh activation function is used for the zero-centered outputs produced by this function.
5. **Dropout layer:** This layer is added to increase the probability of switching the layers off and in turn producing results that are less biased and are less likely to be overfitted.
6. **Dense layer:** The layer consists of 4 neurons, with the activation function Tanh.
7. **Dense layer:** The layer consists of 2 neurons, with the activation function Relu.
8. **Dense layer:** The layer consists of 1 neuron which represents the 2 possible outputs of the given classes, the positive sentiment, and the negative sentiment.

In this model, binary cross-entropy is used as a loss function which compares each of the predicted probabilities to actual class output which can be either 0 or 1. It then calculates the score that penalizes the probabilities based on the distance from the expected value that is how close or far from the actual value. FTRL (Follow The Regularized Leader) is used as an optimizer which is the most suitable for our large data-centric model.

4.2.4 RECUPERATION RECOMMENDATION PHASE

In this phase, the first step is to integrate the models developed in the previous phase for both Textual data and Fitbit data. The weighted average method is used to combine the results of both the sentimental analysis of textual emotion result along with the mood classification through physical parameters result to get the final mood of the user using Eq. 4.9. In the equation, M_1 and M_2 represent the models developed for mood prediction using textual emotions data and physical parameters data from Fitbit device, respectively. w_1 and w_2 refers to the weights assigned to the predictions of the respective models. The values of the weights are 0.8 and 0.2 for w_1 and w_2, respectively and are decided empirically to yield the best possible result in the sample space.

$$MoodorMentalState = M_1 * w_1 + M_2 * w_2 \tag{4.9}$$

(i) Recommending Behavioral Activations (RBA): After detecting the mood of the user in the MHD phase, appropriate behavioral activations are recommended for the users to change their mood. BA are divided into two groups- positive and negative so that appropriate recommendations may be recommended to the users. In this, works songs are used as behavioral activations as they are all-time favorites for the majority of the people to change the mood. The recommendations are generated using the K-means clustering algorithm. K-Means algorithm [35] is an iterative unsupervised classification algorithm that is used to classify the dataset into different non-overlapping groups, such that every data point belongs to exactly one group. Here k-means algorithm is used to divide the songs into two non-overlapping clusters. K-Means is a centroid-based algorithm that decides cluster centers on a random basis from both of the sentimental values in its first iteration, then it determines the best centers for the subsequent classes, then in the further iterations the algorithm assigns each data points to its closest centroid. The data points closer to the particular k center generates clusters to the corresponding class. The reason for using the k-means algorithm for clustering is not only because it is simple to use but also it is very helpful in giving a warm start to the position of the centroid and its ability to adapt to new examples. The various attributes of the song database used are 'danceability', 'loudness', 'energy' etc., to classify the song into two classes. The classes have songs to be given to the people having depressive/negative moods and recommended to the people having a positive mood.

In the proposed model, songs played by the users are updated in the preferences of the model which helps the model to learn over time and generate improved recommendations in the future.

4.3 EXPERIMENTS AND RESULTS

Meticulous studies are done to assess the performance of MHD and RBA in the proposed hybrid recommendation system. The dataset for physical parameters are downloaded from `http://www.orcasinc.com`. The dataset contained 73440 tuples in one table and corresponding 2000 entries in another table which are combined to create a bigger dataset with a greater number of parameters and their corresponding sentiments associated with them. Finally, the data which is useful for mood prediction is kept and the rest of the columns are dropped. Each data tuple has 7 attributes that are used as input namely heart rate, galvanic skin response, the temperature of the body, calories burnt, date and time, steps. And the target value is of the binary data type where -1 denotes sadness and 1 denotes happiness of the individual. After the training of the model, the cross-validation method is used for testing the model.

For prediction of mental state using textual emotions, data is downloaded from Twitter (currently X). The Sentiment140 dataset (`https://www.kaggle.com/kazanova/sentiment140`) contains 1,600,000 tweets extracted using the Twitter (currently X) API. The tweets are annotated (0 = negative, 2 = neutral, 4 = positive) and that they are often used to detect sentiments. This dataset contains 6 attributes:

1. **Target:** The polarity of the tweet (0 = negative, 2 = neutral, 4 = positive).
2. **ID:** The unique identifier of the tweet (e.g., 2087).
3. **Date:** The timestamp of the tweet (e.g., Sat May 16 23:58:44 UTC 2009).
4. **Flag:** The query associated with the tweet (e.g., `lyx`). If there is no query, this value is set to `NO_QUERY`.
5. **User:** The username of the person who posted the tweet (e.g., `robotic_Killdozer`).
6. **Text:** The content of the tweet (e.g., `"Lyx is cool"`).

The various parameters used to evaluate the performance of the proposed models are precision, recall, F-measure, and accuracy. These are widely used metrics for assessing the effectiveness of classification models. All these parameters are defined using the confusion matrix in Table 4.1.

TABLE 4.1
Confusion Matrix

		Predicted Labels	
	Total Data (P+N)	Positive	Negative
Actual Label	Positive (P)	True Positive (TP)	False Negative (FN)
	Negative (N)	False Positive (FP)	True Negative (TN)

The performance metrics are calculated as follows:

$$\text{Precision (Positive Predictive Value)} = \frac{TP}{TP+FP} \tag{4.10}$$

$$\text{Recall (True Positive Rate)} = \frac{TP}{TP+FN} \tag{4.11}$$

$$\text{F-measure (F1-score)} = \frac{2\text{``Precision``Recall}}{\text{Precision}+\text{Recall}} \tag{4.12}$$

$$\text{Accuracy} = \frac{TP+TN}{TP+FP+TN+FN} \tag{4.13}$$

4.3.1 ANALYSIS OF MOOD DETECTION TECHNIQUES

4.3.1.1 Analysis of Mood Prediction using Textual Emotions

The efficacy of the proposed TF-IDF approach for prediction of the mental state of the subject was compared concerning the popular Bag of Words approach. The BOW is a feature extraction technique used in natural language processing and information retrieval. It is a popular technique used to extract features from the text for developing classification models using machine learning algorithms. This method considers documents simply as a collection of words and ignores the semantics of the text. Therefore, in this approach text is converted into a bag of words which maintains the count of total occurrences of frequently used words in the text. Figures 4.2, 4.3, 4.4, and 4.5 depict the performance of the TF-IDF Technique with respect to BOW for precision, recall, f-score, and accuracy parameters, respectively. The performance of both the algorithms is assessed for varied training and test ratio. It can be observed that TF-IDF gives better performance as compared to BOW algorithm on different training test ratio. The best value of precision, F-Score, and accuracy is obtained for 80% training and 20% testing data. However, recall is best on 60% training data.

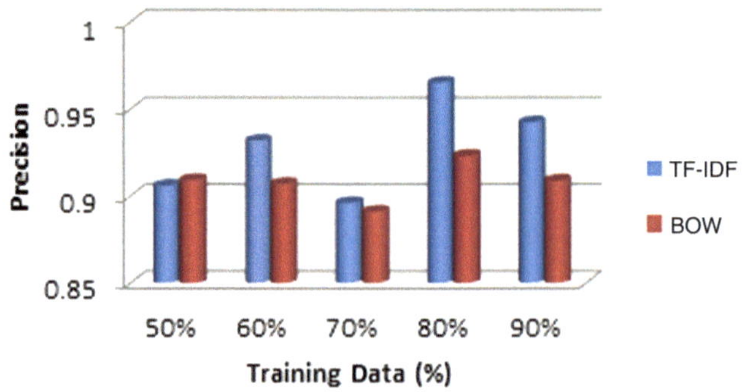

FIGURE 4.2 Precision of TF-IDF vs BOW.

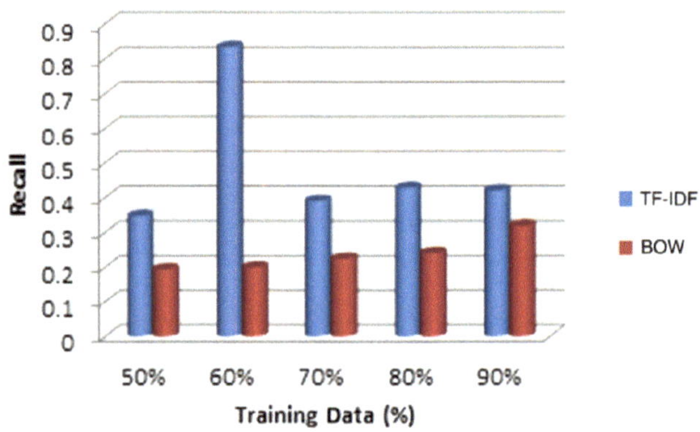

FIGURE 4.3 Recall of TF-IDF vs BOW.

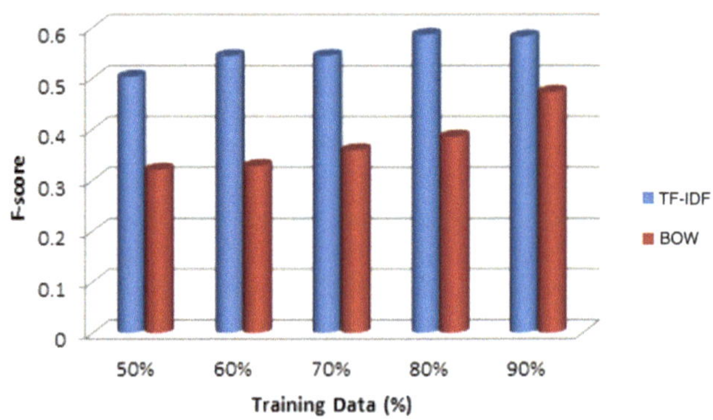

FIGURE 4.4 F-Score of TF-IDF vs BOW.

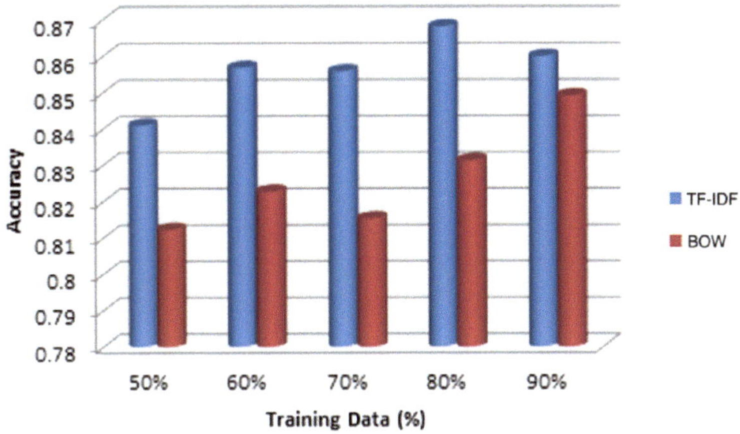

FIGURE 4.5 Accuracy of TF-IDF vs BOW.

4.3.1.2 Mood Classification using Physical parameters Fitbit Data

The performance of the prediction models is determined by the various attributes chosen as physical parameters. For a good prediction model, all the chosen attributes should be independent of each other. To analyze the importance of chosen attributes, a correlation matrix is computed as shown in Figure 4.6. The correlation matrix is a measure to find the correlation between each pair of attributes in a data set. The high correlation value between attributes indicates strong relation between attributes. The lesser correlation between two attributes is highly favorable to any machine learning model's efficiency because if the magnitude of the correlation between attributes is high its efficiency gets impacted due to the multi-collinearity problem. Figure 4.6

	hr	steps	gsr	calories	temp
hr	1.000	0.374	0.075	0.828	-0.111
steps	0.374	1.000	-0.001	0.471	-0.258
gsr	0.075	-0.001	1.000	0.072	0.029
calories	0.828	0.471	0.072	1.000	-0.194
temp	-0.111	-0.258	0.029	-0.194	1.000

FIGURE 4.6 Correlation matrix of physical parameters.

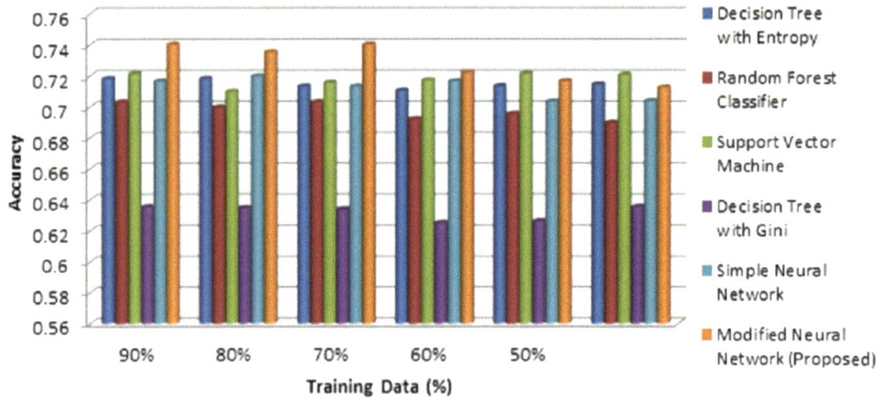

FIGURE 4.7 Mental health detection using machine learning models.

shows that the correlation between the two attributes is fairly low, which shows the quality of the dataset is very good for the model.

Figure 4.7 depicts the prediction accuracy of the Mental health detection model using physical parameters of Fitbit device data. The proposed modified Neural Network algorithm used for training the prediction model is compared with respect to existing supervised machine learning algorithms. The various compared algorithms are Random Forest, Support Vector Machine, decision tree with entropy as well as Gini index, Simple Neural networks. Results of all algorithms are shown for models developed using the varied percentage of training data.

It can be observed from the results shown in Figures 4.2, 4.3, 4.4, and 4.5 that the precision of TF-IDF algorithms is 4-7% better than the bag of words (BOW) technique. For recall, TF-IDF performs significantly better than BOW by upto 20%.F-score value of TF-IDF is improved by 24% as compared to the BOW method. F-sore represents the harmonic balance between both precision and recall which is highest for 80% of training data. Accuracy is also highest when 80% of data is used to train the model. All these results establish that TF-IDF algorithms are the better choice for developing the model and to predict the mental state of users using textual emotional data. Thus, TF-IDF approach is used to develop the model for the proposed hybrid recovery recommender system. It also observed from the results of physical parameters of Fitbit device data that the prediction accuracy of the modified neural network is better than that of all considered machine learning models. Accuracy is improved by upto 10% as compared to all algorithms. Thus, a modified neural network algorithm is used to prepare the recovery recommendation system along with the TF-IDF model discussed in the Section 4.2.3.1.

4.4 DISCUSSION

Since there is no standard method to test the quality of the recommendation system in the real world, the final assessment and analysis of the proposed system are done by amalgamating feedback of the behavioral activations or songs recommended by it. As discussed earlier for testing, 100 persons agreed to share their personal Twitter (currently X) Id along with their physical parameters from the Fitbit band and then made the recommendations. In turn, all these users listened to recommended songs and then made feedback on whether or not the songs were able to help them. Feedbacks were collected on a scale of 1-5 with 1 the least and 5 is the best possible recommendation. Feedbacks have shown that the recommendations are promising and that the recommendation system is working well. 30% of the recommendations were on an exceptional scale while 50% of the results were above average while the rest was just below average and the average rating of the recommendations is 3.25. From these results, it can be deduced that most of the recommendations were up to the mark for the users while some recommendations were not up to the mark as they may or may not have been of the user's taste. Since this system is not very mature and is in the learning process for the particular user, therefore, there might be some recommendations that do not qualify the benchmark. In future this work may be extended to include feedback from the users to further establish the proposed methodology.

4.5 CONCLUSION

In today's world, mental health problems are on a rise and due to arise of the COVID-19 pandemic it's being affected more. Mental health problems need to be focused on. Hence there is a need for applications or ways to help people keep a track of their mental health and also to help them in their dire times. A person who is laughing all the time might be broken from inside and they might crave something or someone who could help them overcome this situation and get back to being their best. For solving such situations this paper presented mental health detection as well as its recuperation through recommendation systems. The work is divided into 2 phases. In the first phase, data is being collected from Twitter (currently X) and Fitbit device. This data is preprocessed using various machine learning techniques. The work also proposed modified neural network algorithm for modeling Fitbit device data. Finally Behavioral activations for the users are generated by developing the mathematical model for integrating the inputs from Twitter (currently X) and Fitbit device. Feedback from the users establish that majority of the recommendations were up to the mark for the users. As the presented system is not very mature and is in the learning process for the particular user, there might be some recommendations that do not qualify the benchmark. With time the results will be improved after every usage by the user. As the proposed systems learn user preferences after every recommendation, it would help in recommending the enhanced behavioral activations for the users in the future.

REFERENCES

1. National Clearinghouse for Mental Health Information (US). *Career Opportunities for Professionals at the National Institute of Mental Health.* Number 2080. US Department of Health, Education, and Welfare, Public Health Service ..., 1970.
2. Nina Helen Mjsund, Monica Eriksson, Irene Norheim, Corey LM Keyes, Geir Arild Espnes, and Hege Forbech Vinje. Mental health as perceived by persons with mental disorders–an interpretative phenomenological analysis study. *International Journal of Mental Health Promotion*, 17(4):215–233, 2015.
3. Sanoar Hossain, Saiyed Umer, Ranjeet Kumar Rout, and Hasan Al Marzouqi. A deep quantum convolutional neural network based facial expression recognition for mental health analysis. *IEEE Transactions on Neural Systems and Rehabilitation Engineering*, 2024.
4. Jakob E Bardram and Aleksandar Matic. A decade of ubiquitous computing research in mental health. *IEEE Pervasive Computing*, 19(1):62–72, 2020.
5. Pim Cuijpers. Four decades of outcome research on psychotherapies for adult depression: An overview of a series of meta-analyses. *Canadian Psychology/psychologie canadienne*, 58(1):7, 2017.
6. Sona Dimidjian, Steven D Hollon, Keith S Dobson, Karen B Schmaling, Robert J Kohlenberg, Michael E Addis, Robert Gallop, Joseph B McGlinchey, David K Markley, Jackie K Gollan, et al. Randomized trial of behavioral activation, cognitive therapy, and antidepressant medication in the acute treatment of adults with major depression. *Journal of consulting and clinical psychology*, 74(4):658, 2006.
7. Darius A Rohani, Andrea Quemada Lopategui, Nanna Tuxen, Maria Faurholt-Jepsen, Lars V Kessing, and Jakob E Bardram. Mubs: A personalized recommender system for behavioral activation in mental health. In *Proceedings of the 2020 CHI Conference on Human Factors in Computing Systems*, pages 1–13, 2020.
8. Mark Deady, David Johnston, David Milne, Nick Glozier, Dorian Peters, Rafael Calvo, Samuel Harvey, et al. Preliminary effectiveness of a smartphone app to reduce depressive symptoms in the workplace: feasibility and acceptability study. *JMIR mHealth and uHealth*, 6(12):e11661, 2018.
9. Matthew Fuller-Tyszkiewicz, Ben Richardson, Britt Klein, Helen Skouteris, Helen Christensen, David Austin, David Castle, Cathrine Mihalopoulos, Renee O'Donnell, Lilani Arulkadacham, et al. A mobile app–based intervention for depression: End-user and expert usability testing study. *JMIR mental health*, 5(3):e9445, 2018.
10. Kien Hoa Ly, Anna Truschel, Linnea Jarl, Susanna Magnusson, Tove Windahl, Robert Johansson, Per Carlbring, and Gerhard Andersson. Behavioural activation versus mindfulness-based guided self-help treatment administered through a smartphone application: a randomised controlled trial. *BMJ open*, 4(1):e003440, 2014.
11. Darius A Rohani, Nanna Tuxen, Andrea Quemada Lopategui, Maria Faurholt-Jepsen, Lars V Kessing, and Jakob E Bardram. Personalizing mental health: a feasibility study of a mobile behavioral activation tool for depressed patients. In *Proceedings of the 13th EAI International Conference on Pervasive Computing Technologies for Healthcare*, pages 282–291, 2019.
12. Fabian Wahle, Tobias Kowatsch, Elgar Fleisch, Michael Rufer, Steffi Weidt, et al. Mobile sensing and support for people with depression: a pilot trial in the wild. *JMIR mHealth and uHealth*, 4(3):e5960, 2016.
13. Adam Mikus, Mark Hoogendoorn, Artur Rocha, Joao Gama, Jeroen Ruwaard, and Heleen Riper. Predicting short term mood developments among depressed patients using

adherence and ecological momentary assessment data. *Internet interventions*, 12:105–110, 2018.
14. Alexandra Cernian, Adriana Olteanu, Dorin Carstoiu, and Cristina Mares. Mood detector-on using machine learning to identify moods and emotions. In *2017 21st International Conference on Control Systems and Computer Science (CSCS)*, pages 213–216. IEEE, 2017.
15. Monika Khandelwal and Ranjeet Kumar Rout. Prmxai: protein arginine methylation sites prediction based on amino acid spatial distribution using explainable artificial intelligence. *BMC bioinformatics*, 24(1):376, 2023.
16. Monika Khandelwal and Ranjeet Kumar Rout. Deepprms: advanced deep learning model to predict protein arginine methylation sites. *Briefings in Functional Genomics*, page elae001, 2024.
17. Ranjeet Kumar Rout, Saiyed Umer, Monika Khandelwal, Smitarani Pati, Saurav Mallik, Bunil Kumar Balabantaray, and Hong Qin. Identification of discriminant features from stationary pattern of nucleotide bases and their application to essential gene classification. *Frontiers in Genetics*, 14:1154120, 2023.
18. Sourish Ghosh, Anasuya Dasgupta, and Aleena Swetapadma. A study on support vector machine based linear and non-linear pattern classification. In *2019 International Conference on Intelligent Sustainable Systems (ICISS)*, pages 24–28. IEEE, 2019.
19. Shasha Mo and Jianwei Niu. A novel method based on ompgw method for feature extraction in automatic music mood classification. *IEEE Transactions on Affective Computing*, 10(3):313–324, 2017.
20. Mahta Bakhshizadeh, Ali Moeini, Mina Latifi, and Maryam Tayefeh Mahmoudi. Automated mood based music playlist generation by clustering the audio features. In *2019 9th International Conference on Computer and Knowledge Engineering (ICCKE)*, pages 231–237. IEEE, 2019.
21. Mariangela Sciandra and Irene Carola Spera. A model-based approach to spotify data analysis: a beta glmm. *Journal of Applied Statistics*, 49(1):214–229, 2022.
22. Alexandros Zenonos, Aftab Khan, Georgios Kalogridis, Stefanos Vatsikas, Tim Lewis, and Mahesh Sooriyabandara. Healthyoffice: Mood recognition at work using smartphones and wearable sensors. In *2016 IEEE international conference on pervasive computing and communication workshops (PerCom Workshops)*, pages 1–6. IEEE, 2016.
23. Bruce Ferwerda and Markus Schedl. Personality-based user modeling for music recommender systems. In *Machine Learning and Knowledge Discovery in Databases: European Conference, ECML PKDD 2016, Riva del Garda, Italy, September 19-23, 2016, Proceedings, Part III 16*, pages 254–257. Springer, 2016.
24. Parul Hgarwal, Mayank Sourabh, Rishabh Sachdeva, Siddharh Sharma, and Shikha Mehta. Recommending optimal tour for groups using ant colony optimization. In *2018 Eleventh International Conference on Contemporary Computing (IC3)*, pages 1–6. IEEE, 2018.
25. Zsolt Mezei and Carsten Eickhoff. Evaluating music recommender systems for groups. *arXiv preprint arXiv:1707.09790*, 2017.
26. Vinothini Kasinathan, Aida Mustapha, Tan Sau Tong, MFCA Rani, and NAA Rahman. Heartbeats: music recommendation system with fuzzy inference engine. *Indonesian J. Electric. Eng. Comput. Sci*, 16:275–282, 2019.
27. Yuchen Dong, Xiaotong Guo, and Yuchen Gu. Music recommendation system based on fusion deep learning models. In *Journal of Physics: Conference Series*, volume 1544, page 012029. IOP Publishing, 2020.

28. Markus Schedl, Hamed Zamani, Ching-Wei Chen, Yashar Deldjoo, and Mehdi Elahi. Current challenges and visions in music recommender systems research. *International Journal of Multimedia Information Retrieval*, 7:95–116, 2018.
29. T Yazawa, AM Hutapea, Y Otomura, K Sagawa, Y Yamada, T Suzuki, and T Katsuyama. Measurement of stress using dfa heartbeat analysis. In *Proceedings of the World Congress on Engineering and Computer Science*, volume 2, 2013.
30. Song Qin, Ronaldo Menezes, and Marius Silaghi. A recommender system for youtube based on its network of reviewers. In *2010 IEEE Second International Conference on Social Computing*, pages 323–328. IEEE, 2010.
31. Saiyed Umer and Ranjeet Kumar Rout. Descriptive and inferential analysis of features for dysphonia and dysarthria parkinson's disease symptoms. *Health Services and Outcomes Research Methodology*, pages 1–23, 2023.
32. Anay Ghosh, Bibhas Chandra Dhara, Chiara Pero, and Saiyed Umer. A multimodal sentiment analysis system for recognizing person aggressiveness in pain based on textual and visual information. *Journal of Ambient Intelligence and Humanized Computing*, 14(4):4489–4501, 2023.
33. Quan Yuan, Gao Cong, and Chin-Yew Lin. Com: a generative model for group recommendation. In *Proceedings of the 20th ACM SIGKDD international conference on Knowledge discovery and data mining*, pages 163–172, 2014.
34. Sang-Woon Kim and Joon-Min Gil. Research paper classification systems based on tf-idf and lda schemes. *Human-centric Computing and Information Sciences*, 9:1–21, 2019.
35. Jure Leskovec, Anand Rajaraman, and Jeffrey David Ullman. *Mining of massive data sets*. Cambridge university press, 2020.

5 Role of Artificial Intelligence in Detection of Congenital Diseases

Kamalpreet Kaur and Renu Dhir

5.1 INTRODUCTION

Congenital infections are diseases or illnesses caused by an organism that poisons a pregnant woman, spreads through the placenta, and moves inside the fetus through the bloodstream. This highly impacts both pregnant women and fetuses. Once the mother gets infected from the initial stage of pregnancy, it continues till childbirth. An impact on neurodevelopment outcomes affects mental health. The mother will affect the child through the placenta, which is known as a Transplacental infection. Also, the infection that passes to the baby during delivery time is known as a peripartum infection. Both infections affect mental health. There are various diseases detected during pregnancy in the neonatal stage or later infancy after delivery like Sepsis, multi-organ failure, Herpes simplex virus (HSV), Varicella zoster virus (VZV), Cytomegalovirus (CMV), Asphyxia, chromosome abnormalities (CA), Zika virus, Parvovirus, Varicella (chickenpox), congenital heart disease (CHD), streptococci, premature delivery, and perinatal death which results in disorder of functional, structural, metabolic, and behavioural abnormalities [1].

These illnesses have the potential to harm every organ in the body. Comprehending the genesis of these disorders is crucial for their eradication in genetic counselling and prevention. Although the exact cause of birth abnormalities is unknown, it is thought to be complex. Of them, 10–30% can be attributed to genetics, 20–35% to multifactorial inheritance, 5–10% to environmental influences, and approximately 30–40% remain unclear. Medical past reveals that important environmental factors are infectious agents, including maternal reasons like age, history, pregnancy illness, medication practice, régime, prenatal care and inappropriate taking of periconceptional folic acid.

Inheritable congenital disorders, parental consanguinity and stillbirths, and prior miscarriages are essential factors in the aetiology of congenital diseases. Salmonella spp., Campylobacter fetus, Listeria monocytogenes, and T. gondii are among the

DOI: 10.1201/9781032660714-5

bacteria that should not be eaten raw or undercooked, including pork, beef, and lamb. Additionally, if a pregnant woman's partner has genital herpes or HIV, she should refrain from having sex. The reduction of risk in infants contracting HIV after birth and breastfeeding should not be made by mothers who are symptomatic or have low CD4 T cell levels [2].

Numerous studies indicate that there is a very significant chance of misdiagnosing it as a result of inadequate research and sporadic pregnancy follow-up. Artificial Intelligence-based applications have been a blessing for premature babies and newborns so that they survive longer through early identification, monitoring, and diagnosis of congenital disorders. Additionally, machine learning has been crucial in a variety of medical research and computational biology fields [3–6]. There has even been an improvement in the congenital disease prediction results [7]. The framework of this chapter provides a flow of predicting congenital disease in Section 2, details various congenital diseases in Section 3, and Section 4 provides several algorithms that have already been used to forecast different congenital illnesses. Lastly, Section 5 includes a conclusion.

5.2 FLOW OF PREDICTING CONGENITAL DISEASE

The congenital disease has been identified from the dataset after the cleaning, amending, and classification process shown in Figure 5.1.

Following the initial phase, users, sites, or sources may provide qualitative or quantitative data for experiments using pertinent data [8, 9]. The experimental data collection results are a procedural process that gathers and examines the dataset. The necessary data will then be chosen based on some available data. In resolving the current problem, any superfluous data, such as replica, disturbance, and lost data, may be removed from this stage. It's possible to stop the occurrences from happening [8, 10].

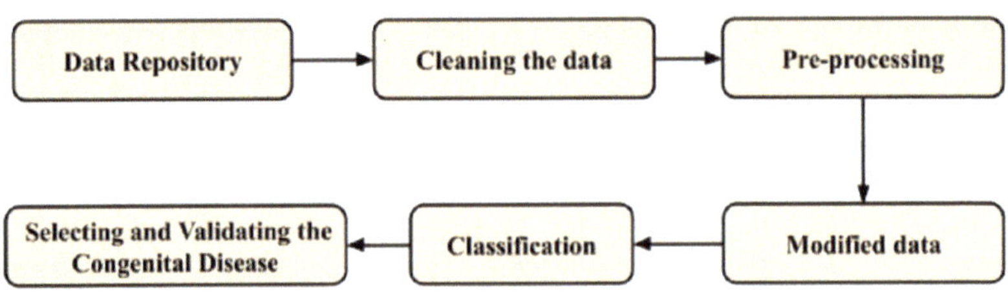

FIGURE 5.1 Process for diagnosis of congenital disease.

5.3 CATEGORIZATION OF CONGENITAL DISEASE OR DISORDER

Numerous congenital illnesses can manifest as postnatal, prenatal, or neonatal symptoms. The most commonly affected diseases like Zika Virus, Varicella-Zoster Virus (Chickenpox), Chlamydia, IUGR, Hepatitis Sepsis, Streptococcus CMV, and CHD with their ailment to newborns, an etiology and its etiologic agents are addressed in Table 5.1 and Figure 5.2. The infection results in the fetal birth and results into the death. If the baby is born premature, it doesn't have good health.

5.4 ALGORITHM FORECAST DIFFERENT CONGENITAL ILLNESSES

5.4.1 FETAL HEALTH PREDICTION

Pregnancy-related fatalities and the rise in associated problems have made it a primary worldwide health concern these days. Every day, problems associated with pregnancy and childbirth account for around 830 maternal fatalities worldwide. This mainly happened in developing nations. Haghpanahi et al. [11] researched fetal subspaces iterative decomposition performance. The complexity, such as ventricular depolarization in non-invasive fetal, is overcome through an Electrocardiogram (ECG) signal by implanting four electrodes on the mother's abdomen. Abbasi et al. [12] suggested that the algorithm uses a Quality Report Program (QRS) that merges components to identify fetal health. Datasets like fetal electrocardiogram datasets (Physio Net/CinC) face challenges during the system training and evaluation. In clinical practice, the results are fetched for various morphological signals and strengths after processing an algorithm. Miao et al. [13] diagnosis of fetal diseases using a multiclass morphologic pattern of the deep learning-based model—artificial intelligence-based method for cardiotocographic diagnosis of the fetus by targeting into classes.

Davidst et al. [14] researched congenital malformations using the DNN (Deep Neural Network) model. The model produces an efficient result with performance parameters like Accuracy, Precision, Recall, and F-score values. Akbulut et al. [15] suggested a system with the help of an electronic -health application for expectant mothers and practitioners. They used a Decision Forest, Logistic Regression (LR), Support Vector Machine (SVM) binary classification models, Neural Network (NN), Bayes Point Machine, Decision Tree, and Boosted DT. Pregnant women, 96 in number, were used in the clinical dataset to train all of the above models. As an output, these applications can provide information about the health state of expectant mothers, consider clinical history data, propose physical activities for expectant mothers, and alert patients to any anomalies.

TABLE 5.1
Various Congenital Diseases with Reasons and Their Images

Disease	Reasons	Etiologic Agents	Image
Cytomegalovirus (CMV) infections	Transmission through contact with contaminated saliva, semen, urine, and vaginal fluid	Virus	Figure 5.2(a)
Chlamydia	Affects both men and women and is sexually transmitted	Bacteria	Figure 5.2(b)
Hepatitis B	Occurs upon contact with the blood and bodily fluids of an infected individual	Virus	Figure 5.2(c)
Herpes simplex virus	One out of every five pregnant women has this infection	Virus	Figure 5.2(d)
Parvovirus B19	Occurs when a child is exposed to parvovirus B19	Virus	Figure 5.2(e)
Rubella (German measles)	Transmitted through the bloodstream and bodily secretions such as the throat and nose	Virus	Figure 5.2(f)
Pediatric syphilis	Sexually transmitted infection passed from a pregnant mother to her unborn child during birth	Bacteria	Figure 5.2(g)
Varicella-Zoster Virus (Chickenpox)	A contagious disease caused by the varicella-zoster virus	Virus	Figure 5.2(h)
Zika Virus	Transmitted through mosquito bites and sexual contact with an infected individual	Virus	Figure 5.2(i)
Intrauterine growth restriction (IUGR)	Caused by viruses, bacteria, fungi, and parasites	Infection	Figure 5.2(j)
Sepsis	Caused by bacteria, viruses, parasites, and fungi	Infection	Figure 5.2(k)
Streptococcus	Transmitted through close contact with an infected individual's mucous or sores	Bacteria	Figure 5.2(l)
Congenital heart disease	Structural issues in the baby's heart, including arteries, veins, or blood flow abnormalities	Disease	Figure 5.2(m)

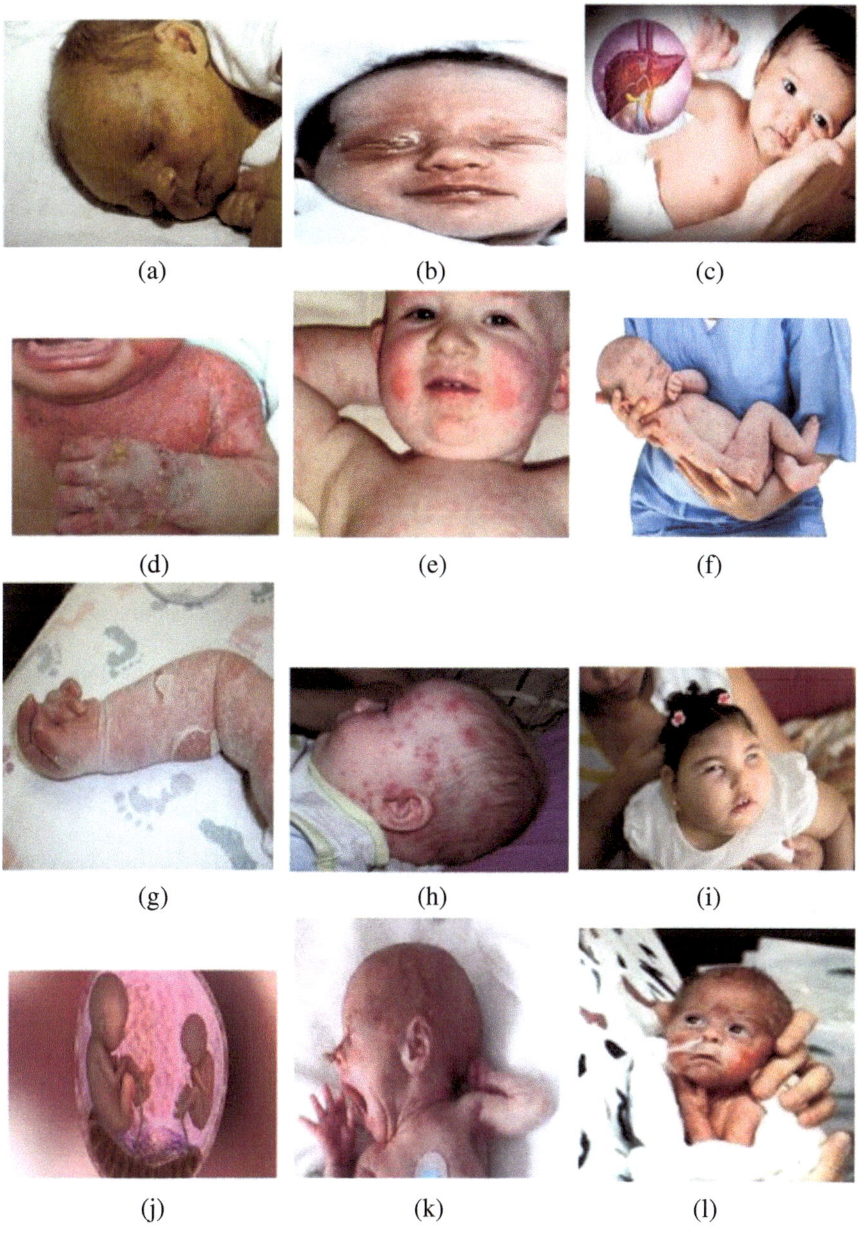

FIGURE 5.2 Demonstration of images for congenital diseases consider under this study. Refer to Table 5.1 for descriptions for (a)–(m).

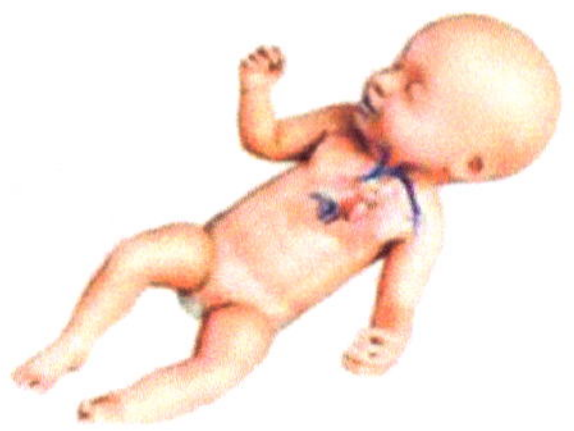

(m)

FIGURE 5.2 (*Continued*) Demonstration of images for congenital diseases consider under this study.

5.4.2 CHROMOSOME ANOMALIES PREDICTION

Feng et al. [16] designed a prediction and screening model for accurate results using deep learning techniques. The model diagnoses the Down syndrome. Neocleous et al. [17] collected the clinical genotyping data for creating a chromosome Single Nucleotide polymorphism (SNP) map that can attain an average accuracy of 99.3%. The researchers have put forth an approach to training on a dataset of Singleton pregnancy with 122362 euploid and 967 aneuploid cases. The training using ANN on raw marker value from mother and fetus [18]. The classification approach to estimate the risk of fetal trisomy 21, along with chromosomal abnormalities, showed an accuracy of 99.5% and 97% of sensitivity. Qin et al. [19] reported the deep convolutional neural network to immediately categorise chromosomal type and polarity. The proposed technique worked on 1909 karyotyping images and results with an accuracy 99.2%. Jaganathan et al. [9] worked on lower-cost prenatal diagnostics to detect genetic conditions using machine learning.

5.4.3 CONGENITAL HEART DISEASES (CHD) PREDICTION

In the world, the most prevalent birth disorders are congenital heart disease (CHD). Numerous studies have examined its risk variables, but none have assessed its predictive power. Luo et al. [20] designed a machine-learning model to validate the risk of carrying infants with congenital heart diseases. Li et al. [21] surveyed congenital disabilities in six Chinese countries. The researchers employed an artificial neural network model to predict the risk that occurred during congestive heart failure in the expected mother. They succeeded in obtaining a 78% sensitivity, 90% specificity, and 68% Yuden Index. According to the research that is currently available, about 50% of newborns have CHD that is not diagnosed at birth. Accordingly, Yoon et al. [22] used Echo to diagnose CHD incidence identification and systemically reviewed the data based on the recommendations of this patient group. Du et al. [23] trained a model with two-level residents in the dataset of 68,969 patients to identify congenital heart disease (CHD). The literature has been reviewed on 630 studies that used ANCM on 1928 newborns to diagnose CHD. The experiment's results, from the model with 94.07% specificity and 74.73% sensitivity and, was able to attain 92.30% accuracy.

Arnaout et al. [24] have worked on a dataset with 107,823 pictures from 1326 retrospective echocardiograms from screening ultrasounds from fetuses between 18 and 24 weeks. With its help, an ensemble neural network was trained to identify the difference between simple and complex conduction heart disease. Its trial yielded a 99% accuracy rate with a 95% sensitivity and specificity. Additionally, Nurmaini et al. [25] presented mechanized fetal heart echocardiogram evaluations based on deep learning models. Additionally, they intrinsically segmented the four conventional heart images and used the instance segmentation approach to identify abnormalities.

5.4.4 INTRA-UTERINE GROWTH RETARDATION (IUGR) PREDICTION

Intra-uterine Growth Retardation (IUGR) is a syndrome that can cause due to numerous reasons and results in morbidity and death of fetuses and neonates. The reason for IUGR is unknown. However, the issue identified more accurately is the Fetal Origin of Adult Disease nowadays. Buscema et al. [26] characterized the relevant insulin-like growth factor using the IUGR dataset, which includes 46 participants categorized into 14 characteristics. Wosiak et al. [27] predicted fetal growth factors using two Neural Network-based innovative algorithms. An automated machine learning-based model predicts effective results. Bahado-Singh et al. [28] used a classifier with combination rules through plurality vote, except simple majority and unanimity. The pathophysiology of IUGR was examined, and artificial intelligence (AI) approaches were used on a variety of platforms, including mass spectrometry (MS)-based metabolomic analysis and nuclear magnetic resonance (NMR) spectroscopy. Support Vector Machines are perfect for attaining accurate results compared to other algorithms. Sufriyana et al. [29] created Placental dysfunction-related illnesses such as IUGR and preeclampsia involve a referral conclusion to be made within a specific timeframe. As a result, suitable prediction models for these illnesses are required for PDDs using machine learning, which comprises parameters related to gestational age.

5.4.5 CYTOMEGALOVIRUS (CMV) PREDICTION

Cytomegalovirus (CMV) is a congenital infection agent for a frequent congenital form of hyperacusis for non-genetic sensorineural hearing loss. Arav-Boger et al. [30] built an ANN model and used it on two datasets to assess the viability of forecasting congenital CMV illness. The experiment's findings indicate 80% and 85% gene prediction outcomes, respectively. Tanimura et al. [31] reviewed the state of knowledge about possible biomarkers for congenital CMV infection prediction. Positive pregnant women (CMV immunoglobulin M (IgM)-positive) are considered at risk for congenital CMV infection if CMV DNA is detected in the mother's uterine cervical secretions. Rogers et al. [32] described numerous experimental experiences using a CMV T cell Immunity Panel (CMV-TCIP), including a test for assessing D4+ and CD8+ T cell responses and a prediction tool. Deepa et al. [33] processed a CMV image to obtain better results from the impacted photographs to capitalize on the benefits of digital image processing in enhancing image quality. The samples from the different patients were examined to obtain an accuracy of 66% for CD8+ T cells and 78% for CD4+ T cells using this method. Hu et al. [34] proposed the deep convolutional neural network for fashion cytometry data analysis. The algorithm has been processed using PCA reduction, ROI and MAFE algorithms. The HSVM has been employed to yield the best classification results. Lee et al. [35] analyzed the model using a permutation-based method, and the results indicate that the population of CD27–, CD94+, and CD8+ T cells was identified as being strongly related to latent CMV infection by the suggested model. Eisenberg et al. [36] proposed a machine learning model for integrating data of patients within time-dependent laboratory measures. The mortality has been predicted and reactivated during the hematopoietic cell transplantation (HCT) periods per patient.

5.4.6 ZIKA VIRUS PREDICTION

Positive-sense RNA flavivirus, known as the Zika virus (ZIKV), is spread mainly by the Aedes aegypti mosquito. Jiang et al. [37] proposed a backward propagation neural network (BPNN) with RF and GBM to find the congenital severe anomalies known as the Zika pandemic. The researchers carried out fifty modelling operations using the training dataset to assess the performance of machine learning models. The results indicate that BPNN yields the best results with 95% accuracy [38]. For training, Lusk et al. [39] used machine learning algorithms like LR, GBT SVM, RF, KNN, and DT and gathered the appropriate classification results. Herry et al. [40] hypothesised that heart rate variability (HRV) in ZIKV-affected children might provide a biomarker of fetal ZIKV exposure. Veiga et al. [10] proposed and unveiled a machine-learning system for testing the theory in young infants exposed to the Zika virus during pregnancy. They employed a cubic SVM classifier for evaluation, providing 92% negative predictive value and 86% positive predictive value that identify ZIKV patients from structured and unstructured public data. Dadheech et al. [41] used an algorithm to test positive for ZIKV infection, underwent an RNA test and reached 83% accuracy when identifying data without textual information from medical records and 76% accuracy regarding image reports.

5.4.7 SEPSIS PREDICTION

Masino et al. [42] created a model employing quick electronic health record (EHR) data that could identify baby sepsis at least four hours before it was identified clinically. The method that models and improves the results of Sepsis. Song et al. [43] investigated the model's viability using machine learning and non-invasive vital sign data and evaluated it using electronic medical records. About 85–87% accuracy is reported in the results, with no significant variations. Quiros et al. [44] used various patient monitoring signals 24 hours before onset to predict late-onset Sepsis in preterm infants. It was discovered that the AUC of the 48-hour prediction model for the late-onset neonatal sepsis algorithm is 86.1%, and for the onset detection model, it was 86.8%. Kopanitsa et al. [45] explored a machine learning perspective for sepsis prediction and risk management by identifying sepsis risk in the early stages of pregnancy using real-world evidence data. The model was tested using patient electronic health information from the Almazov Perinatal Medical Centre with an AUC of 95%.

5.4.8 PREECLAMPSIA (PE) HYPERTENSIVE PREDICTION

Preeclampsia, also known as PE, is an irreversible hypertensive pregnancy condition. However, research indicates that early detection of PE can avoid it. Using maternal heart rate variability indices, Tejera et al. [46] constructed an artificial neural network (ANN) model to classify pregnant women with preeclamptic and hypertension conditions at different gestational ages. The results obtained with the proposed model indicate that PE increases the standard and hypertensive pregnancy groups with a sensitivity of 80% and a specificity of 85–90%. Moreira et al. [47] developed a further neural fuzzy machine learning strategy for the prediction of HELLP syndrome, a complex hypertensive disorder that occurs during pregnancy. Later, Tahir et al. [48] employed the NN and DL algorithms to enhance risk prediction related to PE levels in pregnant women. They employed 17 parameters from 1077 patient records from two Makassar hospitals and Haji General Surabaya Hospital, collected between Dec. 12, 2017, and Feb. 12, 2018, to analyze the suggested method. The necessary features were then chosen using PSO, and the result shows an accuracy of 95.12%. Bennett et al. [49] used cost-sensitive DNN to predict PE one year later, considering data with significant imbalances and racial disparities.

5.5 CONCLUSION

This study highlights the pivotal role of artificial intelligence-based models in detecting congenital diseases in pregnant women. By analyzing medical imaging data from MRIs, echocardiograms, and ultrasounds, AI-powered algorithms can identify subtle abnormalities—such as cardiac defects, skeletal irregularities, and neural tube malformations—that may indicate congenital conditions. Furthermore, AI-driven predictive models leverage historical patient data, family history, and environmental factors to assess the likelihood of congenital disorders. These models empower healthcare providers to prioritize early interventions by categorizing patients based on risk levels. In conclusion, the integration of AI into prenatal diagnostics has significantly enhanced the ability of medical professionals to anticipate congenital conditions, enabling more informed decision-making and guiding future research. Despite the notable progress in AI-driven diagnostics, challenges persist in achieving high diagnostic accuracy. Addressing these obstacles is crucial to refining detection methods and ensuring effective treatment for emerging congenital diseases.

REFERENCES

1. Matthew H Spitzer and Garry P Nolan. Mass cytometry: single cells, many features. *Cell*, 165(4):780–791, 2016.
2. Meherwar Fatima, Maruf Pasha, et al. Survey of machine learning algorithms for disease diagnostic. *Journal of Intelligent Learning Systems and Applications*, 9(01):1, 2017.
3. Monika Khandelwal and Ranjeet Kumar Rout. Prmxai: protein arginine methylation sites prediction based on amino acid spatial distribution using explainable artificial intelligence. *BMC bioinformatics*, 24(1):376, 2023.
4. Monika Khandelwal, Ranjeet Kumar Rout, and Saiyed Umer. Prediction of protein-protein interaction using support vector machine based on spatial distribution of amino acids. In *International Conference on Advances and Applications of Artificial Intelligence and Machine Learning*, pages 23–32. Springer, 2022.
5. Ranjeet Kumar Rout, Saiyed Umer, Monika Khandelwal, Smitarani Pati, Saurav Mallik, Bunil Kumar Balabantaray, and Hong Qin. Identification of discriminant features from stationary pattern of nucleotide bases and their application to essential gene classification. *Frontiers in Genetics*, 14:1154120, 2023.
6. Monika Khandelwal, Nazir Shabbir, and Saiyed Umer. Extraction of sequence-based features for prediction of methylation sites in protein sequences. In *Artificial intelligence technologies for computational biology*, pages 29–46. CRC Press, 2022.
7. James B McAuley. Congenital toxoplasmosis. *Journal of the Pediatric Infectious Diseases Society*, 3(suppl˙1):S30–S35, 2014.
8. Yue Huang, Han Wang, Tingdong Li, Caihong Li, Jiabao Tang, Huan Yu, Xiaoyi Guo, Qiaoqiao Song, Feixue Wei, Jiangding Wang, et al. Comparison of detection strategies for screening and confirming congenital cytomegalovirus infection in newborns in a highly seroprevalent population: a mother-child cohort study. *The Lancet Regional Health–Western Pacific*, 12, 2021.
9. M Jaganathan, R Gopal, and VR Kiruthika. Modelling an effectual feature selection approach for predicting down syndrome using machine learning approaches. *International Journal of Aquatic Science*, pages 1238–1249, 2019.
10. Rafael V Veiga, Lavinia Schuler-Faccini, Giovanny VA França, Roberto FS Andrade, Maria Glória Teixeira, Larissa C Costa, Enny S Paixão, Maria da Conceição N Costa, Maurício L Barreto, Juliane F Oliveira, et al. Classification algorithm for congenital zika syndrome: characterizations, diagnosis and validation. *Scientific Reports*, 11(1):6770, 2021.
11. Masoumeh Haghpanahi and David A Borkholder. Fetal qrs extraction from abdominal recordings via model-based signal processing and intelligent signal merging. *Physiological measurement*, 35(8):1591, 2014.
12. Hamid Abbasi, Laura Bennet, Alistair J Gunn, and Charles P Unsworth. Robust wavelet stabilized 'footprints of uncertainty'for fuzzy system classifiers to automatically detect sharp waves in the eeg after hypoxia ischemia. *International journal of neural systems*, 27(03):1650051, 2017.
13. Julia H Miao and Kathleen H Miao. Cardiotocographic diagnosis of fetal health based on multiclass morphologic pattern predictions using deep learning classification. *International Journal of Advanced Computer Science and Applications*, 9(5), 2018.
14. Adeniji Oluwashola Davidst, Omopintemi Ayorinde Henry, and Samuel Oladele Adeyemi. Prediction of congenital anomaly using deep learning techniques. In *2023 First International Conference on the Advancements of Artificial Intelligence in African Context (AAIAC)*, pages 1–4. IEEE, 2023.

15. Akhan Akbulut, Egemen Ertugrul, and Varol Topcu. Fetal health status prediction based on maternal clinical history using machine learning techniques. *Computer methods and programs in biomedicine*, 163:87–100, 2018.
16. Bing Feng, David C Samuels, William Hoskins, Yan Guo, Yan Zhang, Jijun Tang, and Zibo Meng. Down syndrome prediction/screening model based on deep learning and illumina genotyping array. In *2017 IEEE International Conference on Bioinformatics and Biomedicine (BIBM)*, pages 347–352. IEEE, 2017.
17. AC Neocleous, A Syngelaki, KH Nicolaides, and Christos N Schizas. Two-stage approach for risk estimation of fetal trisomy 21 and other aneuploidies using computational intelligence systems. *Ultrasound in Obstetrics & Gynecology*, 51(4):503–508, 2018.
18. Andreas C Neocleous, Kypros H Nicolaides, and Christos N Schizas. Intelligent noninvasive diagnosis of aneuploidy: raw values and highly imbalanced dataset. *IEEE Journal of biomedical and health informatics*, 21(5):1271–1279, 2016.
19. Yulei Qin, Juan Wen, Hao Zheng, Xiaolin Huang, Jie Yang, Ning Song, Yue-Min Zhu, Lingqian Wu, and Guang-Zhong Yang. Varifocal-net: A chromosome classification approach using deep convolutional networks. *IEEE transactions on medical imaging*, 38(11):2569–2581, 2019.
20. Yanhong Luo, Zhi Li, Husheng Guo, Hongyan Cao, Chunying Song, Xingping Guo, and Yanbo Zhang. Predicting congenital heart defects: A comparison of three data mining methods. *PloS one*, 12(5):e0177811, 2017.
21. Huixia Li, Miyang Luo, Jianfei Zheng, Jiayou Luo, Rong Zeng, Na Feng, Qiyun Du, and Junqun Fang. An artificial neural network prediction model of congenital heart disease based on risk factors: A hospital-based case-control study. *Medicine*, 96(6):e6090, 2017.
22. Shin Ae Yoon, Woi Hyun Hong, and Hwa Jin Cho. Congenital heart disease diagnosed with echocardiogram in newborns with asymptomatic cardiac murmurs: a systematic review. *BMC pediatrics*, 20:1–10, 2020.
23. Yunmei Du, Shuai Huang, Canhui Huang, Allam Maalla, and Huiying Liang. Recognition of child congenital heart disease using electrocardiogram based on residual of residual network. In *2020 IEEE International Conference on Progress in Informatics and Computing (PIC)*, pages 145–148. IEEE, 2020.
24. Rima Arnaout, Lara Curran, Yili Zhao, Jami C Levine, Erin Chinn, and Anita J Moon-Grady. An ensemble of neural networks provides expert-level prenatal detection of complex congenital heart disease. *Nature medicine*, 27(5):882–891, 2021.
25. Siti Nurmaini, Muhammad Naufal Rachmatullah, Ade Iriani Sapitri, Annisa Darmawahyuni, Bambang Tutuko, Firdaus Firdaus, Radiyati Umi Partan, and Nuswil Bernolian. Deep learning-based computer-aided fetal echocardiography: application to heart standard view segmentation for congenital heart defects detection. *Sensors*, 21(23):8007, 2021.
26. Massimo Buscema, Enzo Grossi, Luisa Montanini, and Maria E Street. Data mining of determinants of intrauterine growth retardation revisited using novel algorithms generating semantic maps and prototypical discriminating variable profiles. *PLoS One*, 10(7):e0126020, 2015.
27. Agnieszka Wosiak, Agata Zamecznik, and Katarzyna Niewiadomska-Jarosik. Supervised and unsupervised machine learning for improved identification of intrauterine growth restriction types. In *2016 Federated Conference on Computer Science and Information Systems (FedCSIS)*, pages 323–329. IEEE, 2016.

28. Ray Oliver Bahado-Singh, Ali Yilmaz, Halil Bisgin, Onur Turkoglu, Praveen Kumar, Eric Sherman, Andrew Mrazik, Anthony Odibo, and Stewart F Graham. Artificial intelligence and the analysis of multi-platform metabolomics data for the detection of intrauterine growth restriction. *PloS one*, 14(4):e0214121, 2019.
29. Herdiantri Sufriyana, Yu-Wei Wu, Emily Chia-Yu Su, et al. Prediction of preeclampsia and intrauterine growth restriction: development of machine learning models on a prospective cohort. *JMIR medical informatics*, 8(5):e15411, 2020.
30. Ravit Arav-Boger, Yuval S Boger, Charles B Foster, and Zvi Boger. The use of artificial neural networks in prediction of congenital cmv outcome from sequence data. *Bioinformatics and Biology Insights*, 2:BBI–S764, 2008.
31. Kenji Tanimura and Hideto Yamada. Potential biomarkers for predicting congenital cytomegalovirus infection. *International journal of molecular sciences*, 19(12):3760, 2018.
32. Ralph Rogers, Kapil Saharia, Aditya Chandorkar, Zoe F Weiss, Kendra Vieira, Sophia Koo, and Dimitrios Farmakiotis. Clinical experience with a novel assay measuring cytomegalovirus (cmv)-specific cd4+ and cd8+ t-cell immunity by flow cytometry and intracellular cytokine staining to predict clinically significant cmv events. *BMC infectious diseases*, 20:1–11, 2020.
33. K Deepa and S Suganya. Multiple attribute feature extraction and high support vector classifier for identification of cytomegalovirus images. *ICTACT Journal on Image & Video Processing*, 11(2), 2020.
34. Zicheng Hu, Alice Tang, Jaiveer Singh, Sanchita Bhattacharya, and Atul J Butte. A robust and interpretable end-to-end deep learning model for cytometry data. *Proceedings of the National Academy of Sciences*, 117(35):21373–21380, 2020.
35. Jung Su Lee, Jihye Yun, Sungwon Ham, Hyunjung Park, Hyunsu Lee, Jeongseok Kim, Jeong-Sik Byeon, Hwoon-Yong Jung, Namkug Kim, and Do Hoon Kim. Machine learning approach for differentiating cytomegalovirus esophagitis from herpes simplex virus esophagitis. *Scientific Reports*, 11(1):3672, 2021.
36. Lisa Eisenberg, XplOit consortium, Christian Brossette, Jochen Rauch, Andrea Grandjean, Hellmut Ottinger, Jurgen Rissland, Ulf Schwarz, Norbert Graf, Dietrich W Beelen, et al. Time-dependent prediction of mortality and cytomegalovirus reactivation after allogeneic hematopoietic cell transplantation using machine learning. *American Journal of Hematology*, 97(10):1309–1323, 2022.
37. Dong Jiang, Mengmeng Hao, Fangyu Ding, Jingying Fu, and Meng Li. Mapping the transmission risk of zika virus using machine learning models. *Acta tropica*, 185:391–399, 2018.
38. B Mahalakshmi and G Suseendran. Prediction of zika virus by multilayer perceptron neural network (mlpnn) using cloud. *Int J Recent Technol Eng (IJRTE)*, 8:1–6, 2019.
39. Richard Lusk, John Zimmerman, Kelley VanMaldeghem, Suzanna Kim, Nicole M Roth, James Lavinder, Anna Fulton, Meghan Raycraft, Sascha R Ellington, and Romeo R Galang. Exploratory analysis of machine learning approaches for surveillance of zika-associated birth defects. *Birth defects research*, 112(18):1450–1460, 2020.
40. Christophe L Herry, Helena MF Soares, Lavinia Schuler-Faccini, and Martin G Frasch. Machine learning model on heart rate variability metrics identifies asymptomatic toddlers exposed to zika virus during pregnancy. *Physiological Measurement*, 42(5):055008, 2021.
41. Pankaj Dadheech, Abolfazl Mehbodniya, Shivam Tiwari, Sarvesh Kumar, Pooja Singh, Sweta Gupta, and Henry kwame Atiglah. Zika virus prediction using ai-driven technol-

ogy and hybrid optimization algorithm in healthcare. *Journal of Healthcare engineering*, 2022(1):2793850, 2022.

42. Aaron J Masino, Mary Catherine Harris, Daniel Forsyth, Svetlana Ostapenko, Lakshmi Srinivasan, Christopher P Bonafide, Fran Balamuth, Melissa Schmatz, and Robert W Grundmeier. Machine learning models for early sepsis recognition in the neonatal intensive care unit using readily available electronic health record data. *PloS one*, 14(2):e0212665, 2019.
43. Wongeun Song, Se Young Jung, Hyunyoung Baek, Chang Won Choi, Young Hwa Jung, Sooyoung Yoo, et al. A predictive model based on machine learning for the early detection of late-onset neonatal sepsis: development and observational study. *JMIR medical informatics*, 8(7):e15965, 2020.
44. Laura Cabrera-Quiros, Deedee Kommers, Maria K Wolvers, Laurien Oosterwijk, Niek Arents, Jacqueline van der Sluijs-Bens, Eduardus JE Cottaar, Peter Andriessen, and Carola van Pul. Prediction of late-onset sepsis in preterm infants using monitoring signals and machine learning. *Critical care explorations*, 3(1):e0302, 2021.
45. Georgy Kopanitsa, Oleg Metsker, David Paskoshev, and Sofia Greschischeva. Identification of risk factors and prediction of sepsis in pregnancy using machine learning methods. *Procedia Computer Science*, 193:393–401, 2021.
46. Eduardo Tejera, Maria Jose areias, Ana Rodrigues, Ana Ramoa, Jose Manuel nieto villar, and Irene Rebelo. Artificial neural network for normal, hypertensive, and preeclamptic pregnancy classification using maternal heart rate variability indexes. *The Journal of Maternal-Fetal & Neonatal Medicine*, 24(9):1147–1151, 2011.
47. Mário WL Moreira, Joel JPC Rodrigues, Jalal Al-Muhtadi, Valery V Korotaev, and Victor Hugo C de Albuquerque. Neuro-fuzzy model for hellp syndrome prediction in mobile cloud computing environments. *Concurrency and computation: Practice and experience*, 33(7):1–1, 2021.
48. Muhlis Tahir, Tessy Badriyah, and Iwan Syarif. Classification algorithms of maternal risk detection for preeclampsia with hypertension during pregnancy using particle swarm optimization. *EMITTER International Journal of Engineering Technology*, 6(2):236–253, 2018.
49. Rachel Bennett, Zuber D Mulla, Pavan Parikh, Alisse Hauspurg, and Talayeh Razzaghi. An imbalance-aware deep neural network for early prediction of preeclampsia. *Plos one*, 17(4):e0266042, 2022.

6 A Hybrid Multi-Level Segmentation-Based Ensemble Classification Model for Multi-Class Diabetic Retinopathy Detection

Shaik Akbar

6.1 INTRODUCTION

Diabetes Mellitus is a metabolic disorder with excess blood glucose characterized by hyperglycemia [1]. It is also referred to as diabetes. Insulin is the pancreas hormone that regulates blood glucose levels in the body. The diabetes-affected person's pancreas cannot either produce or misuse enough insulin, and it can either be used. Insulin allows the body to use carbohydrate glucose in the diet and makes it easier to enter into the body cells to burn energy. The body delicately regulates fasting blood glucose between 4 and 5.5 mm/litre in the plasma. Insulin helps avoid too high (hyperglycemia) or low blood glucose levels (hypoglycemia). Prolonged high levels of glucose in the blood can result in the glycosylation of proteins, which is a significant contributing factor to the long-term complications of diabetes. Diabetes is generally divided into two primary types: Type I, where insulin is absent, and Type II, where there is insulin resistance and inadequate insulin production.

Insulin insufficiency is the essential component of diabetes, either complete or partial. Retinal images can identify and diagnose various eye-related disorders, and human beings play a crucial role. To determine the existence of diabetic retinopathy (DR) lesions in fundus images, a universal logical method was developed [2], which can distinguish red and bright lesions and is free of preprocessing or post-processing requirements. Different steps are taken, and the colored retinal images examine various stages involving diabetic retinopathy. Microa-aneurysms are thin, secular bags

DOI: 10.1201/9781032660714-6

due to the local swelling of the capillary walls and are presented as little red dots. Another method [3] claims the walls are slim and crack simply because they cause bleeding. Hardened exudates are deposits of yellow lipids seen as bright yellow lesions.

The bright circular region of the blood vessel is known as the optic disk. The fovea describes the middle of the retina and is the highest visual acuity area. In diabetic retinopathy, the retinal retina of the eye is weakened in patients with diabetes. This leads to a complete loss of vision. Diabetes is triggered by the body's failure to store and use blood sugar. Early visible signs of the disease typically do not appear, and microaneurysms can be seen as the disease progresses, both soft and strong, exudates and new blood vessels. Median filtering is used to decrease the noise effect. A median filter of 56x56 [4] applies the shade-corrected image on the green plane. Using a linear structure of 25 pixels, the top hat transform is spaced apart by 180 grades. The retinal image is flushed through the kernels, and the variations between the gluey images are determined by thresholding. The image can be normalized and corrected. A 9 x9 mask is used as smoothing. The vessel ends are not detected. The outcome is a threshold, but the candidates on the vessels are not identified, and a duration ratio excludes fake positives. [5] used the neural network, the histogram equalization and the specialization preprocessing level.

The regions covered by the window are used for input purposes using a sliding window technique. The application of the filter algorithm aims to detect the kernel micro-aneurysms, which produce the maximum response. Multi-scale Gaussian filters [6] and wavelet filters [7] are commonly used filter filters. The study results [8], which was done with the segmentation of exudates and image process using the C-Clustering of Fuzzy C by optimizing several clusters, proposed the new approach to protecting protein classifications based on a neural network and feature selection. The Naive Bayes Classifier classification approach has limited research on automated exudate detection. Similarly, [9] for retinal damage detection using K-means and C-Means Algorithm. Sandhu et al. [10] used a HEIMED public database data set, using three analyzes: probability map exudation, colour analysis and wavelets analysis. The findings showed some noise accompanying the exudate analysis. Further studies are necessary to advance the detection and analysis of exudates. Researchers use max tree filtering and noise reduction attributes. The problem is corrected using this study's microneurysms features; exudates are divided into Max trees and mathematical morphology [11].

In addition, before using it for classification, the extraction of features in every local characteristic is selected in advance. The function of the extraction of features is to choose the results with a significant class differential and ignore the non-dominant function extraction results. The variance coefficient can be used to select these features. Saifuddin introduced a new way of detecting exudates [12], in which lesions were classified, then exudates, and morphological operations detected no exudates. With a reported precision of 85%, Sargunar classified the gravity of DR based on its perimeter, region and hurt coefficient. Mahendran and others used the grey grade matric for the exudates extracted, which were detected in morphologic processing. The exudate detection system, blood vessels, and microaneurysm were presented in [13]. The entropy and homogeneity with reported accuracy were calculated over 92% for classification. Identifying blood vessels, exudates, and microaneurysms, as mentioned in [14], forms the basis for classifying retinal images as diabetic or non-diabetic. As indicated in [15], the DR detection and the classification were based on exudates and blood vessels alone: normal and DR. In [16], only blood vessel features and haemorrhages were extracted. Microaneurysms and exudates were not taken into account. Precisely determining the mild, moderate, and severe stages of DR poses a significant challenge, as it is not a straightforward process. In blood vessels and exudates, roughly diabetic or non-diabetic images were detected and classified.

Medical image is a technique and process for producing visual representations for clinical analysis and medical intervention of the internal body part. Medical imaging has become a key tool in clinical trials as visualization and quantitative assessment can be quickly diagnosed. Medical imaging also provides a normal anatomical and physiological database that identifies abnormalities. Visualization of internal structures in the body today is provided by several imaging devices. Medical imaging X-ray methods include conventional X-rays, CTs (Computed Tomography) and mammography. Contrasting agents, such as angiography examinations, enhance the X-ray image. Molecular imaging is used in nuclear medicine and employs several methods for visualising biological processes in organism cells. For molecular imagery, small amounts of radiation-active markers are known as radiopharmaceuticals. Analysis of images requires various steps; the first essential step is to analyze a digital image following preprocessing to identify whether the retina is normal or abnormal by extracting features that distinguish between them by analyzing the retina. Their method used the fluent logic of a pixel group supporting vector machine, a computer vector composed of a histogram of intensity and a grey level based on extraction features for pixel representation. Then, a result according to the evaluation's sensitivity, specificity and accuracy is estimated at 96%. In the method of extracting qualities, the researcher relied on a number of qualities and thus worked fully on the next step (the stage of classification). The classification process takes a great deal of time, affecting the measurement of system precision and improvement in performance. Their method is based on segmentation by the visual disk and extracting blood vibrations according to the histogram using median filters. They propose an algorithm for classifying and detecting (DR) (Pillar-K means algorithm). Their system uses the Pillar.

Optimized K-mean Algorithm to enhance the image in all color spaces to improve accuracy and computing time. [2] the morphological operators were optimally analyzed for the effective detection by non-dilated pupils and low contrast retinal images of microaneurysm. A novel approach to retinal vessel extraction, using mathematical form study and spatial subjective clustering methods, is presented to overcome the disadvantage of diabetes retinopathy detection. The automated identification of retinal images streamlines the process of diagnosing and screening for diabetic retinopathy. Image segmentation involves the utilization of fuzzy clustering and neural networks. This method is a failure in some bright regions, which causes incorrect image segmentation and the resulting regions become too light. [17] proposed the detection of Diabetic Retinopathy with the aid of color fundus images. When the characteristic features are extracted from raw images, they are then entered into the Support Vector Machine with Fuzzy C-means clustering by using an image processing technique. This Fuzzy C-means clustering method integrates SVM (Support Vector Machine) and preprocessing techniques to enhance the detection of blood vessels and optic discs. It employs a hybrid approach for the analysis and mitigation of diabetic retinopathy. The problem with k-means that the local minimum is minimized. The Fuzzy C-means clustering Algorithm is detected in different diabetic retinopathy stages. DR screening helps to detect the symptoms quickly and easily. These modules perform image segmentation, including vessel segmentation, exudate, MA (Micro Aneurysms) and other modules for classifying fluffy c-means. This has Diabetic Retinopathy sensitivity, precision and precision. [18] discusses using colored retinal images for scientific purposes for diabetic retinopathy (DR) and its routine detection. A novel technique is used to recognize MA images from digital color images. The authors also concentrated on successive repudiatory strategies for a less aggressive response. A new approach to MA detection from Digital Color Fundus images is also available. MA detection is considered as an objective disorder detection problem, with a considerably lower probability of a target occurrence than the disorder. A consecutive rejection policy will gradually decrease the number of cluttering reactions.

6.2 LITERATURE SURVEY

Prompt identification of the disease is vital for the efficient care of DR. Those with diabetes are commonly instructed to schedule an annual eye exam with an eye specialist. In the very early phase, this will certainly diagnose DR. The diagnosis of DR involves the detection of lesions and abnormalities in the blood vessels of the retinal fundus image. Utilizing a fundus camera, color images of the retinal background are captured. The student is generally dilated with eye drops when capturing the fundus images. The general approach for image extraction consists of several phases: a preprocessing phase for improvement of contrast or equalization of non-uniformity, Image segmentation, extraction of functions, and classification. The strategy for feature extraction varies by target or landmark for retinal lesions. The guidelines on retinal image analysis were outlined in [19]. Initial techniques to detect retinal landmarks and DR lesions were examined. In recent reviews of disease detection models [20–25] and red lesions [26], the main features (non-deep learning) based approaches in literature are presented focus on DR CAD (computer-aided detection or diagnosis). These techniques have usually shown that interdependence with lesion detection involves the detection of anatomical structures and, in turn, determines an automated DR test result [27, 28].

Poor performance concerning typical anatomical structures could harm lesion detection and screening precision. In the domain of medical image analysis, CNNs (ConvNets) are considered the most favoured option among the different methodological variants of deep learning [29, 29–33]. In [34], the Convolutional Neural Networks from scratch were trained and compared to the fine-tuning findings based on the other two architectures existing. The operator of the felling is similar to what was described but performed in the frequency field, which is to multiply the mask and the image in the frequency domain. This operator combines linear low-pass filters with high-pass frequency filters. Small transmission values define a low-pass filter for high spatial frequencies, while large transfer values define low spatial frequencies. A filter with a low pass has smoothing behaviour with a standard average, and its sharp borders are blurred. On the other hand, a high pass filter is defined by sizeable spatial frequency transfers, while small spatial frequencies transfer. A filter for high passes improves borders. Combining these two filters, a high-pass filter could enhance the edge of the low-intensity, and a low-pass filter could smooth the high-intensity edge. Recently, age and society-related diseases such as diabetes have increased in Sweden and other parts of the world. During the PDR (Proliferative Diabetic Retinopathy) phase, circulatory problems lead to oxygen deprivation or chemical parts of the retina. The vessels are new and fragile as the circulatory system tries to keep the appropriate oxygen levels inside the retina. Neo-vascularization is this phenomenon. It is also the entry and the place for nerves into and out of the retina.

The macula, a structure shaped like an oval, is located near the central area of the retina. The fovea, which contains a high concentration of cone cells, is near the macula's centre. Abnormalities related to the eye can be classified into two main categories: the first category encompasses cataracts, conjunctivitis, blue tongue, and glaucoma, all of which have an impact on the eye. Secondly, lifestyle diseases, including high blood pressure(BP), arteriosclerosis, and diabetes [35] are categorized as lifestyle-related conditions. The sensitivity refers to the percentage of an abnormal fundus image classified by a method as abnormal. In contrast, the particularity of the fundus image can be defined as a normal percentage. The first threshold-based method and region growing are simple, but it isn't easy to select threshold values, seed points, and stop criteria in the image zone. The survey includes these techniques in 2 categories: The first category is based on clustering methods [36], which makes determining the number of clusters the main difficulty with these methods. Mathematical morphology is the basis for the second category, which depends on image contrast. In addition to the previous categories, the researchers use combined approaches that take advantage of two or more categories. Most researchers in this field agreed unanimously to use the preprocessing step as a main step in preparing the segmentation and detection area of interest for the abnormal object based on obtaining an adequate image for feature extraction. [37] proposed an algorithm comprising several operators' reduction of shade variation and exuding their shine and sharpening based on a global threshold value to improve contrast based on their lightness or grey level.

In [38], retina images are a good summary and examination of many automatic exudate detection works that contain acquisition elements to be removed. Many studies show that approximately 12% of retinal images cannot be clinically analyzed for lack of quality [39]. For easy exudate detection, the image preprocessing of the retina is necessary. Image noise can be randomized to information about the brightness or color or visible transfer of the image's grain. This is highly unwanted, particularly in cases in which the characteristics of the exudates are comparable to those produced by noise. Thus, retinal image preprocessing for easily detecting exudates is essential to improve the image quality. The steps involved in preprocessing include i) a green color extraction from RGB (red, green and blue) images and an image recorder that allows the retinal disc to be isolated. (ii) The filtering of an image to remove an artefact because of non-uniform lightening and gauze-based noise removal smoothing. A specialist edge detection algorithm may need to be needed, as traditional edge detection techniques may not be able to segment the vessels from the retinal image exactly. Precise segmentation of vasculature is a difficult task for many reasons, such as the noise presence, the low contrast between vessels and the background and vessel width, luminosity and shape differences. Several methods, the matched filer, edge morphology and vessel segmentation, called pixel grading, tracking methods, supervised classification and wavelet transformation, have been proposed for this purpose. Reflecting on the small, strange surface of the soft tissue of the retina is a difficult task to automatically detect the blood vessels from a retinal image [40], given the low contradiction between the vessels and the background and the pathological variation. Several methods were used for the segmentation of the retinal vessel. There are two main categories for classification: rule-based techniques and supervised techniques. Each color image has its images based on the primary color representation, specifically RGB analysis.

The separation of each plane is so separate as to be called planar separation of this type of function. All color images will initially be decomposed into a grey image, so only the desired feature can be dramatically extracted. The green band is extracted from the original retinal image of RGB to monochrome the input image. The green channel offers the best naval-background contrast to the RGB representation, while the blue one offers poor dynamics, and the red channel is the brightest and lowest contrast. Therefore, the blood that contains the retinal layer elements (such as vessels) is best represented, and the green canal is more in contrast. This system must be used on retinal images of various sizes, and the input image needs to be redimensioned. Diabetes is collectively a disease in which a man is suffering from high glucose, either mildly in such a way that the frame is not producing enough insulin now or the cells do not react to insulin. Diabetic retinopathy poses a significant perplexity for those affected by diabetes, making it one of the most frequent concerns in managing the condition. It's a far and spreads the eye disease to most components. It hurts the small veins of the retina and causes a loss of imagination and foresight. The risk of infection increases with age, and diabetic retinopathy is moderately matured and more organized. Early retinopathy is diabetic non-proliferative retinopathy. Minor veins spill blood or fluid into the retina at this level. The fluid discharge causes the retina, known as exudates, to swell and shape. Proliferative diabolic retinopathy, the PDR, is an eye-creating effort to build or resupply the retina with newly selected (neovascular) vessels, regardless of how the retinal blood is delivered.

6.3 PROPOSED FRAMEWORK

The proposed method uses hybrid ensemble feature extraction models as kernels to improve the accuracy of motion vectors in training frames. The feature extraction process is significantly enhanced by integrating a 3x3x3 ensemble kernel estimator as a filter, along with five pooling layers and two fully connected layers. Additionally, the segmentation model is seamlessly integrated into the C3D framework to eliminate any noise that might impact feature identification. It is recommended that an ensemble multi-class SVM within the C3D framework be utilized to predict anomalies based on the extracted features. Automated diagnostic systems for diabetic retinopathy (DR) greatly benefit from segmentation in their development. Threshold-based segmentation is a fundamental preprocessing step in various algorithms for automatically detecting different features. Segmentation is crucial for establishing a reference framework within retinal images, making it essential for any image analysis application. Although several methods for optical disk localization have been introduced, many focus solely on locating the optical disk without addressing contour detection challenges. A binary image containing bright regions can be produced by thresholding the intensity image. By capitalizing on these distinctive characteristics, an advanced classification system driven by deep learning is utilized to identify and classify disease patterns within an extensive training dataset. The process, exemplified in Figure 6.1, is meticulously executed in two stages.

Initially, the image attributes are dissected, and a filter-oriented Bayesian deep neural network is employed to categorize the disease severity by effectively detecting shape characteristics. This initial stage also encompasses the application of an innovative mathematical filter, allowing the detection of shape characteristics and the performance of crucial statistical analysis, which is pivotal to accurately classifying disease severity. This state-of-the-art filter is vital in minimizing noise and smoothening dark red dots and bright spot lesions. Subsequently, a deep CNN network is employed in the second stage to analyze the filtered dataset, aiming to extract essential features that bear significance in the classification process. Throughout this deep learning phase, the features necessary for classification in each training image are further enhanced by employing a `pre-trained` C3D framework. Ultimately, in the classification stage, a novel classifier is introduced to produce predictions regarding disease classes. A hybrid `non-linear` Bayesian SVM classifier is employed to train the C3D features used in disease class prediction to secure the accuracy of classification results.

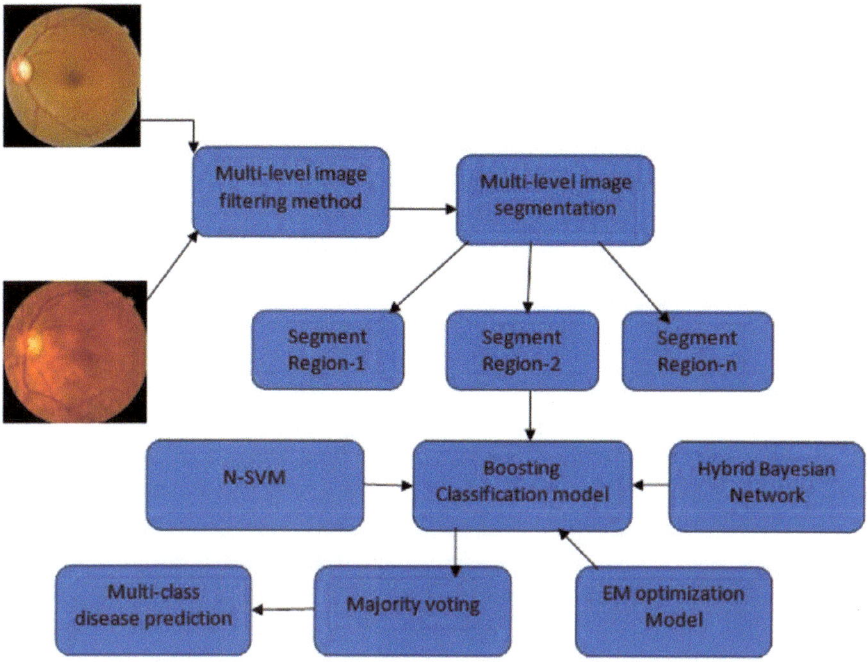

FIGURE 6.1 Proposed segmentation-based Bayesian deep neural network.

6.3.1 THRESHOLD FEATURE SEGMENTATION

Parting each image into blocks allows for identifying the nearest neighbour features and their variations. By denoting B_i and B_j as partitions representing the means and covariance matrices of the i_{th} and j_{th} blocks within a frame, this method involves analyzing each block and its neighboring blocks to detect and annotate disease patterns and their orientations within the image. This is illustrated in Algorithm 6.1.

6.3.2 PROPOSED MULTI-LEVEL FEATURE SELECTION-BASED DEEP LEARNING FRAMEWORK USING BOOSTING CLASSIFIERS

Traditional methods can operate in real-time on low-end systems and perform satisfactorily under specific conditions such as simple contexts, fixed lighting, and similar constraints. However, their efficiency significantly diminishes in more complex environments. Contrastingly, deep learning approaches based on CNNs offer profound capabilities in identifying single or multi-class anomalies. Unlike conventional object detection methods that rely on multiple features to locate objects in images using low-level visual cues, CNNs struggle to capture representative semantic information in intricate scenarios. Moreover, conventional detection pipelines often construct and configure each stage separately, hindering the attainment of an optimal global solution for the entire system. Inspired by the visual cortex's organization in the brain,

Algorithm 6.1 Threshold Feature Segmentation

Input: Training disease and normal images

Filter the images using disease patterns

Perform multi-level segmentation using the thresholding method along with the correlation measure

For each region in the initially segmented regions, compute:

$$Seg(G_{\min}(p,q),r) = \max\left\{\log(r,\lambda), \frac{\exp\left(-p/2\sigma_p^2\right)}{\sigma_p^{2\pi}}, \log(r,\eta), \frac{\exp\left(-q/2\sigma_q^2\right)}{\sigma_q^{2\pi}}\right\} \tag{6.1}$$

$$Seg(G_{\max}(p,q),r) = \max\left\{\log(r,\lambda), \frac{\exp\left(-p/2\sigma_p^2\right)}{\sigma_p^{2\pi}}, \log(r,\eta), \frac{\exp\left(-q/2\sigma_q^2\right)}{\sigma_q^{2\pi}}\right\} \tag{6.2}$$

Define the multi-segmented threshold for over-segmented regions as:

$$MST_1(I,\lambda,\eta) = \max(\lambda,\eta)\sum_{i=1}^{\infty}\left(\frac{\min\left(p\cdot \text{Prob}\frac{I}{c_1}, q\cdot \text{Prob}\frac{I}{c_2}\right)}{N}\right) \tag{6.3}$$

Extract each disease intensity feature value of ROI and apply kernel density estimation. Compute feature correlation based on mutual information and correlation measure:

$$e_k = \frac{1}{2}\text{correlation}(F_0(r))\cdot(MI_0(r) - O_0(r))^2 \tag{6.4}$$

$$O_0(r) = \sum_{k}^{L} MI_{ho}(r)\psi_{ho} \tag{6.5}$$

$$N_{ho}(r) = X(N_{ih}(O_0(r)))\cdot \text{gauss}(N) \tag{6.6}$$

$$N_{ih}(r) = \sum_{i}^{m} w_{ih}x_i + a_h \tag{6.7}$$

$$MSG(p,q) = N_{ho}(r)\exp\left(\frac{p_1^2}{\sigma_p^2}\right)\frac{\log\left(\frac{2\pi}{\lambda}\right)}{e_k} \tag{6.8}$$

$$FS(G,D(p,q),C_m) = \max\left\{\frac{\text{prob}\left(\frac{MGD(p,q)}{C_m}\right)}{MGD(p,q)\,\text{Prob}(C_m)}\right\} \tag{6.9}$$

End

CNN architecture incorporates convolutional layers and fully connected (FC) layers within a neural network framework. Convolutional layers extract spatial features from input data, while FC layers contribute to decision-making based on these extracted features. The proposed model employs max-pooling to down-sample feature maps and reduce computational complexity. This involves dividing input images into non-overlapping two-dimensional regions and selecting the maximum value within each area to create neuron clusters. Consequently, two fully connected layers and an output layer facilitate the classification stage. These interconnected layers establish connections between each neuron in one layer and every neuron in the subsequent layer. They process the weighted sum of features from the previous layer to generate the target output predictions. The input to each fully connected layer is a vector derived from the dimensions of the preceding max-pooling layer and the depth of neurons.

6.3.2.1 Multi-Class Classification Model

A non-linear optimization function has been created for the SVM model to enhance prediction accuracy by reducing the error rate. The boosted CNN-SVM classifier, which includes a non-linear kernel function, is employed on feature sets for class prediction. SVM multi-class optimization models are utilized on each feature set as the proposed classification model for predicting disease severity levels based on C3D features. Deep learning features are initially trained using a feature extraction algorithm and filtered using the C3D model across severity classes. A new non-linear multi-class classification model is presented to forecast disease severity levels within the chosen feature space (described in Algorithm 6.2). In comparison of non-linear multi-class SVM classification model, the Algorithm 6.3 gives the algorithmic steps for the implementation of the Multi-Class Bayesian Network classifier.

Algorithm 6.2 Non-linear Multi-class Classification

1. Feature space FS=C3DF (Filter (Images))
2. To each C3D feature f in FS
3. Apply the non-linear SVM classification on the features using the following objective function and kernel function as the Minimization of the second objective function is represented as

$$\min J = \frac{1}{2}\left(\text{sign}(w \cdot \phi(x)_i + k)\right)^2 + \max(O_1) = \omega \sum_{i=1}^{k} \sum_{j=1}^{RD} \log P_{ij} + \varepsilon v_t \sum_{i=1}^{n} (\gamma_i + \gamma_i) \tag{6.10}$$

$$P_{ij} = \frac{\text{Prob}\left(\frac{g_i}{\text{Tr}[j] \exp\left(\frac{D(g_i, \text{Tr}[j])}{\min(G_t, RD)}\right)}\right)}{\sum_{m=1}^{k} \exp\left(\frac{D(g_i, \text{Tr}[j])}{mT_{rc}}\right)}$$

$$g_i \in G_t, \quad Tr_i \in RD, \quad k: \text{number of segments}$$

$$T_{rc} = \frac{\sum_{i=1}^{Tr} g_i Tr_i}{\sum_{j=1}^{Tr} g_j}$$

$$v_t = \frac{1}{2} \ln \frac{1 - \gamma_t}{\gamma_t}$$

$$\varepsilon = \arg\min v_t e$$

Algorithm 6.3 Multi-Class Bayesian Network

Input: Image I divided into blocks B_i

for each block B_i in I **do**

Compute the normalized Gaussian kernel on block B_i

$$G(B_i, B_j, \sigma_B) = \frac{1}{2\pi\sigma_B^2} \exp\left(\frac{i^2 + j^2}{2\sigma_B^2}\right) \tag{6.11}$$

Compute the angular variation of the block Angular Directional Gradient(ADG):

$$ADG = \tan^{-1}\left(\frac{B_j}{B_i}\right), \quad ADG \in (0, 2\pi] \tag{6.12}$$

Compute histogram intensities:

$$HI_f = \text{HistIntensities}(I) \tag{6.13}$$

Compute the Gaussian Conditional Estimator (GCE):

$$\text{GCE}(\alpha, \beta) = \frac{\exp(\alpha^2)}{\eta \cdot \beta^2} \tag{6.14}$$

Compute α and β:

$$\alpha = GKE\left(\sum HI_f, \sum V_f\right) \tag{6.15}$$

$$\eta : \text{Scaling Factor}, \quad \beta = \text{div}(H(\text{Gr}(V_f))(\text{Gr}(V_f))) \tag{6.16}$$

Compute the Kernel Conditional Probability (KCP):

$$\text{KCP}(D) = \left| \frac{HI_f}{\sum \max(\alpha, \beta) \cos(HI_f)} \right| \tag{6.17}$$

end for

6.4 EXPERIMENTAL RESULTS

Experimental outcomes were simulated using Python and Java environments within a cloud computing framework. The MESSIDOR image dataset, designed explicitly for diabetic retinopathy computer-assisted diagnoses, was utilized in this study `https://www.adcis.net/en/third-party/messidor/`. Some image samples of this database has been demonstrated in Figure 6.2. The outcomes of Algorithm 1 has been demonstrated in Figure 6.3. Researchers are encouraged to use this database to evaluate and improve their algorithms, thereby contributing to advancements in the field. The Messidor database is a valuable resource for the research community, facilitating the testing and refining techniques for diabetic retinopathy

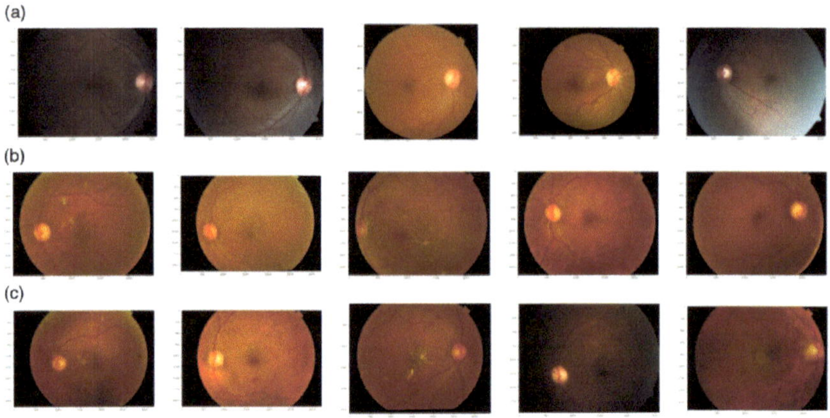

FIGURE 6.2 DR images at different phases of severity: (a) represents no visible signs of the disease (NO DR), (b) represents the presence of small, round dots on the retina (microaneurysms), and (c) represents the presence of any of the subsequent signs, with no evidence of proliferative diabetic retinopathy.

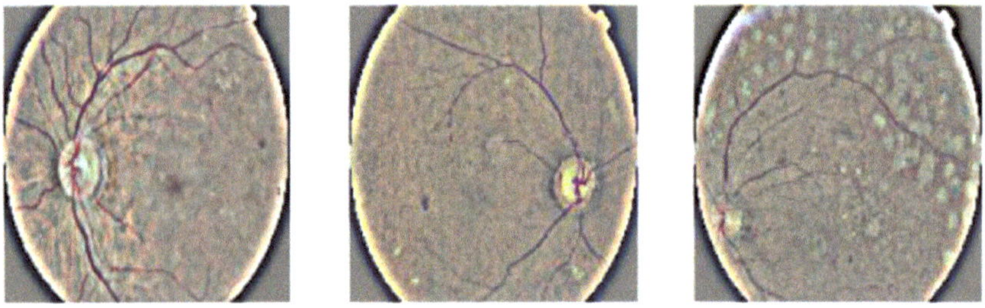

FIGURE 6.3 After preprocessing and feature extraction process.

detection and analysis. Amazon AWS cloud servers with multiple graphic processing unit (GPU) instances were employed to simulate the image dataset. The proposed system's effectiveness was evaluated using metrics such as specificity (Spec), sensitivity (Sen), accuracy (Acc), and positive predictive value (PPV).

$$Sen = \frac{TP}{TP + FN} \tag{6.18}$$

$$Spec = \frac{TN}{TN + FP} \tag{6.19}$$

$$PPV = \frac{TP}{TP+FP} \tag{6.20}$$

$$ACC = \frac{TP+FN}{TP+TN+FP+FN} \tag{6.21}$$

(i) TP (True Positive): Predicting disease patterns in the MA region.
(ii) FP (False Positive): Wrongly predicting patterns in the non-MA region.
(iii) TN (True Negative): Correctly predicting disease patterns in Non-MA regions.
(iv) FN (False Negative): Wrongly predicting patterns in non-MA regions.

Figures 6.4, 6.5, and 6.6 illustrate the recall, precision, and F-measure of the classifiers from Algorithm 2 and Algorithm 3 in comparison with other classifiers.

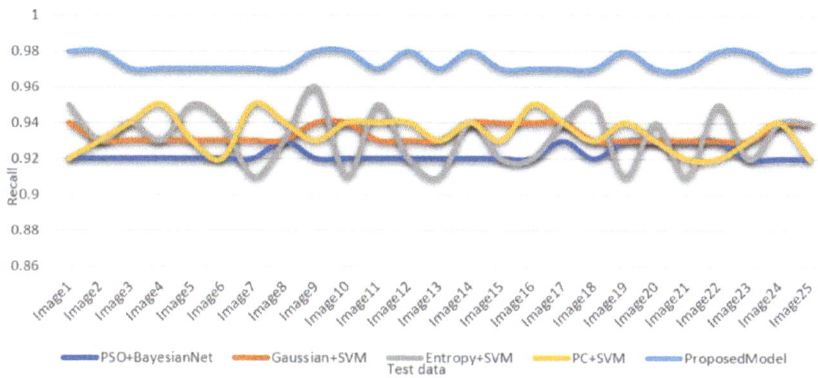

FIGURE 6.4 Comparative result of presented probabilistic multi-disease classification approach to the conventional approaches for recall measure.

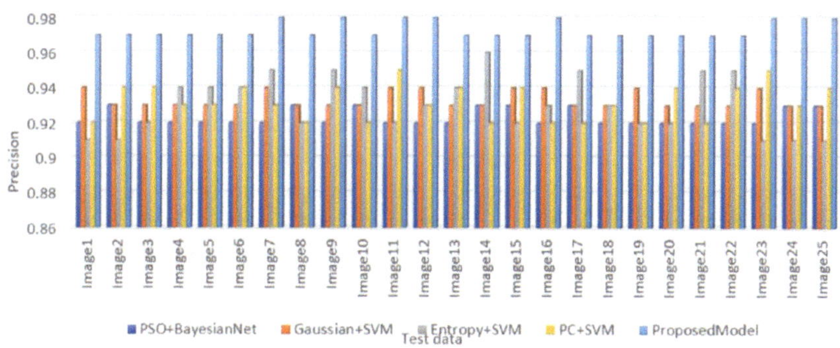

FIGURE 6.5 Comparative result of presented probabilistic multi-disease classification approach to the conventional approaches for precision measure.

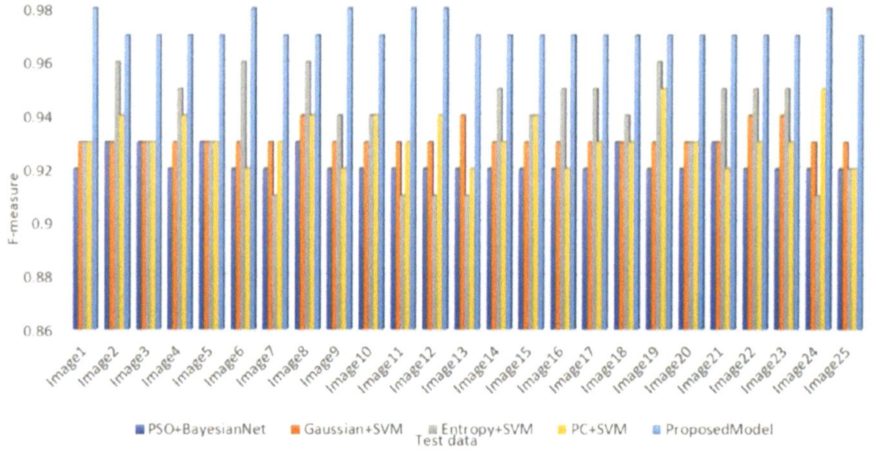

FIGURE 6.6 Comparative result of presented probabilistic multi-disease classification approach to the conventional approaches for F-measure measure.

Additionally, Table 6.1 and Table 6.2 present the comparative performance in terms of accuracy and error rate, respectively.

6.5 CONCLUSION

This chapter implements a hybrid multi-class classification model on the DR multi-class database. Most conventional models are limited to binary class classification problems with limited dimensions and computing resources. Also, these models have high multi-class detection error rates and actual negative rates due to noise and sparsity issues. To solve these multi-class issues, an advanced multi-class segmentation-based classification model is proposed to improve the efficiency of the DR multi-class prediction. This work proposes hybrid expectation maximization (EM) and hybrid ensemble classification techniques for the CNN network. Experimental results proved that the proposed multi-class segmentation-based classification problem has better efficiency (2.5%) improvement on the accuracy, recall, F-measure and precision and 1.2% improvement on the runtime (ms) than the conventional models.

TABLE 6.1

Performance Results of the Proposed Probabilistic Multi-Disease Pattern Detection Accuracy Compared to Conventional Models on the Training Dataset

Test Data	PSO Bayesian Net	Gaussian SVM	Entropy SVM	PC SVM	Proposed Model
Image 1	0.92	0.93	0.91	0.93	0.98
Image 2	0.92	0.93	0.94	0.93	0.97
Image 3	0.92	0.93	0.94	0.94	0.98
Image 4	0.92	0.93	0.91	0.93	0.98
Image 5	0.92	0.93	0.91	0.93	0.98
Image 6	0.92	0.93	0.91	0.93	0.98
Image 7	0.92	0.93	0.91	0.93	0.98
Image 8	0.92	0.93	0.91	0.93	0.98
Image 9	0.92	0.93	0.91	0.93	0.98
Image 10	0.92	0.93	0.91	0.93	0.98
Image 11	0.92	0.93	0.91	0.93	0.98
Image 12	0.92	0.93	0.91	0.93	0.98
Image 13	0.92	0.93	0.91	0.93	0.98
Image 14	0.92	0.93	0.91	0.93	0.98
Image 15	0.92	0.93	0.91	0.93	0.98
Image 16	0.92	0.93	0.91	0.93	0.98
Image 17	0.92	0.93	0.91	0.93	0.98
Image 18	0.92	0.93	0.91	0.93	0.98
Image 19	0.92	0.93	0.91	0.93	0.98
Image 20	0.92	0.93	0.91	0.93	0.98
Image 21	0.92	0.93	0.91	0.93	0.98
Image 22	0.92	0.93	0.91	0.93	0.98
Image 23	0.93	0.94	0.95	0.94	0.98
Image 24	0.92	0.93	0.94	0.93	0.97
Image 25	0.93	0.94	0.92	0.93	0.97

TABLE 6.2
Comparative Result of Presented Probabilistic Multi-Disease Classification Approach to the Conventional Approaches for Error Rate Measure

Test Data	PSO+ BayesianNet	Gaussian+ SVM	Entropy+ SVM	PC+ SVM	Proposed
Image1	0.08	0.07	0.08	0.07	0.03
Image2	0.07	0.06	0.05	0.06	0.03
Image3	0.07	0.06	0.08	0.06	0.03
Image4	0.08	0.07	0.05	0.07	0.03
Image5	0.08	0.07	0.08	0.06	0.03
Image6	0.08	0.07	0.09	0.06	0.03
Image7	0.08	0.07	0.08	0.07	0.02
Image8	0.08	0.06	0.05	0.06	0.03
Image9	0.08	0.07	0.06	0.07	0.03
Image10	0.08	0.07	0.08	0.07	0.03
Image11	0.08	0.07	0.08	0.07	0.02
Image12	0.08	0.07	0.07	0.06	0.03
Image13	0.07	0.07	0.07	0.07	0.02
Image14	0.07	0.07	0.05	0.06	0.03
Image15	0.08	0.07	0.05	0.07	0.03
Image16	0.08	0.06	0.09	0.08	0.02
Image17	0.08	0.07	0.08	0.07	0.03
Image18	0.08	0.07	0.08	0.07	0.03
Image19	0.07	0.06	0.08	0.05	0.02
Image20	0.08	0.07	0.08	0.08	0.03
Image21	0.07	0.06	0.05	0.07	0.02
Image22	0.08	0.07	0.07	0.07	0.03
Image23	0.08	0.07	0.06	0.07	0.03
Image24	0.08	0.07	0.06	0.07	0.03
Image25	0.07	0.07	0.05	0.07	0.03

REFERENCES

1. Jordi de La Torre, Aida Valls, and Domenec Puig. A deep learning interpretable classifier for diabetic retinopathy disease grading. *Neurocomputing*, 396:465–476, 2020.
2. Skylar Stolte and Ruogu Fang. A survey on medical image analysis in diabetic retinopathy. *Medical image analysis*, 64:101742, 2020.
3. Rishi P Singh, Michael J Elman, Simran K Singh, Anne E Fung, and Ivaylo Stoilov. Advances in the treatment of diabetic retinopathy. *Journal of Diabetes and its Complications*, 33(12):107417, 2019.
4. Qiwei Xie, Yanfei Liu, Hui Huang, Bei Hong, Jinxin Wang, Hua Han, and Yue Liu. An innovative method for screening and evaluating the degree of diabetic retinopathy and drug treatment based on artificial intelligence algorithms. *Pharmacological Research*, 159:104986, 2020.
5. Yi-Ting Hsieh, Lee-Ming Chuang, Yi-Der Jiang, Tien-Jyun Chang, Chung-May Yang, Chang-Hao Yang, Li-Wei Chan, Tzu-Yun Kao, Ta-Ching Chen, Hsuan-Chieh Lin, et al. Application of deep learning image assessment software verisee™ for diabetic retinopathy screening. *Journal of the Formosan Medical Association*, 120(1):165–171, 2021.
6. Yuchen Xie, Quang D Nguyen, Haslina Hamzah, Gilbert Lim, Valentina Bellemo, Dinesh V Gunasekeran, Michelle YT Yip, Xin Qi Lee, Wynne Hsu, Mong Li Lee, et al. Artificial intelligence for teleophthalmology-based diabetic retinopathy screening in a national programme: an economic analysis modelling study. *The Lancet Digital Health*, 2(5):e240–e249, 2020.
7. Valentina Bellemo, Zhan W Lim, Gilbert Lim, Quang D Nguyen, Yuchen Xie, Michelle YT Yip, Haslina Hamzah, Jinyi Ho, Xin Q Lee, Wynne Hsu, et al. Artificial intelligence using deep learning to screen for referable and vision-threatening diabetic retinopathy in africa: a clinical validation study. *The Lancet Digital Health*, 1(1):e35–e44, 2019.
8. K Shankar, Abdul Rahaman Wahab Sait, Deepak Gupta, S Kd Lakshmanaprabu, Ashish Khanna, and Hari Mohan Pandey. Automated detection and classification of fundus diabetic retinopathy images using synergic deep learning model. *Pattern Recognition Letters*, 133:210–216, 2020.
9. Abhishek Samanta, Aheli Saha, Suresh Chandra Satapathy, Steven Lawrence Fernandes, and Yu-Dong Zhang. Automated detection of diabetic retinopathy using convolutional neural networks on a small dataset. *Pattern Recognition Letters*, 135:293–298, 2020.
10. Harpal Singh Sandhu, Mohammed Elmogy, Ahmed Taher Sharafeldeen, Mohamed Elsharkawy, Nabila El-Adawy, Ahmed Eltanboly, Ahmed Shalaby, Robert Keynton, and Ayman El-Baz. Automated diagnosis of diabetic retinopathy using clinical biomarkers, optical coherence tomography, and optical coherence tomography angiography. *American journal of ophthalmology*, 216:201–206, 2020.
11. Ashish Bora, Siva Balasubramanian, Boris Babenko, Sunny Virmani, Subhashini Venugopalan, Akinori Mitani, Guilherme de Oliveira Marinho, Jorge Cuadros, Paisan Ruamviboonsuk, Greg S Corrado, et al. Predicting the risk of developing diabetic retinopathy using deep learning. *The Lancet Digital Health*, 3(1):e10–e19, 2021.
12. Wei Zhang, Jie Zhong, Shijun Yang, Zhentao Gao, Junjie Hu, Yuanyuan Chen, and Zhang Yi. Automated identification and grading system of diabetic retinopathy using deep neural networks. *Knowledge-Based Systems*, 175:12–25, 2019.
13. Rishab Gargeya and Theodore Leng. Automated identification of diabetic retinopathy using deep learning. *Ophthalmology*, 124(7):962–969, 2017.

14. Yusaku Katada, Nobuhiro Ozawa, Kanato Masayoshi, Yoshiko Ofuji, Kazuo Tsubota, and Toshihide Kurihara. Automatic screening for diabetic retinopathy in interracial fundus images using artificial intelligence. *Intelligence-Based Medicine*, 3:100024, 2020.
15. Cam-Hao Hua, Thien Huynh-The, Kiyoung Kim, Seung-Young Yu, Thuong Le-Tien, Gwang Hoon Park, Jaehun Bang, Wajahat Ali Khan, Sung-Ho Bae, and Sungyoung Lee. Bimodal learning via trilogy of skip-connection deep networks for diabetic retinopathy risk progression identification. *International Journal of Medical Informatics*, 132:103926, 2019.
16. Norah Asiri, Muhammad Hussain, Fadwa Al Adel, and Nazih Alzaidi. Deep learning based computer-aided diagnosis systems for diabetic retinopathy: A survey. *Artificial intelligence in medicine*, 99:101701, 2019.
17. Xuechen Li, Linlin Shen, Meixiao Shen, Fan Tan, and Connor S Qiu. Deep learning based early stage diabetic retinopathy detection using optical coherence tomography. *Neurocomputing*, 369:134–144, 2019.
18. Recep E Hacisoftaoglu, Mahmut Karakaya, and Ahmed B Sallam. Deep learning frameworks for diabetic retinopathy detection with smartphone-based retinal imaging systems. *Pattern recognition letters*, 135:409–417, 2020.
19. Katrine B Nielsen, Mie L Lautrup, Jakob KH Andersen, Thiusius R Savarimuthu, and Jakob Grauslund. Deep learning–based algorithms in screening of diabetic retinopathy: A systematic review of diagnostic performance. *Ophthalmology Retina*, 3(4):294–304, 2019.
20. Tahira Nazir, Aun Irtaza, Zain Shabbir, Ali Javed, Usman Akram, and Muhammad Tariq Mahmood. Diabetic retinopathy detection through novel tetragonal local octa patterns and extreme learning machines. *Artificial intelligence in medicine*, 99:101695, 2019.
21. Jie Xue, Shuo Yan, Jianhua Qu, Feng Qi, Chenggong Qiu, Hongyan Zhang, Meirong Chen, Tingting Liu, Dengwang Li, and Xiyu Liu. Deep membrane systems for multitask segmentation in diabetic retinopathy. *Knowledge-Based Systems*, 183:104887, 2019.
22. Francisco J Martinez-Murcia, Andrés Ortiz, Javier Ramírez, Juan M Górriz, and Ricardo Cruz. Deep residual transfer learning for automatic diagnosis and grading of diabetic retinopathy. *Neurocomputing*, 452:424–434, 2021.
23. Nasr Y Gharaibeh. Withdrawn: Detection of diabetic retinopathy using partial swarm optimization (pso) and gaussian interval type-2 fuzzy membership functions (git2fmfs), 2020.
24. Wejdan L Alyoubi, Wafaa M Shalash, and Maysoon F Abulkhair. Diabetic retinopathy detection through deep learning techniques: A review. *Informatics in Medicine Unlocked*, 20:100377, 2020.
25. Gabriel Tozatto Zago, Rodrigo Varejão Andreão, Bernadette Dorizzi, and Evandro Ottoni Teatini Salles. Diabetic retinopathy detection using red lesion localization and convolutional neural networks. *Computers in biology and medicine*, 116:103537, 2020.
26. Ankita Gupta and Rita Chhikara. Diabetic retinopathy: Present and past. *Procedia computer science*, 132:1432–1440, 2018.
27. Tao Li, Yingqi Gao, Kai Wang, Song Guo, Hanruo Liu, and Hong Kang. Diagnostic assessment of deep learning algorithms for diabetic retinopathy screening. *Information Sciences*, 501:511–522, 2019.
28. Teresa Araujo, Guilherme Aresta, Luís Mendonça, Susana Penas, Carolina Maia, Ângela Carneiro, Ana Maria Mendonça, and Aurélio Campilho. Dr— graduate: Uncertainty-aware deep learning-based diabetic retinopathy grading in eye fundus images. *Medical Image Analysis*, 63:101715, 2020.

29. T Vijayan, M Sangeetha, A Kumaravel, and B Karthik. Withdrawn: Gabor filter and machine learning based diabetic retinopathy analysis and detection, 2020.
30. Gaurav Saxena, Dhirendra Kumar Verma, Amit Paraye, Alpana Rajan, and Anil Rawat. Improved and robust deep learning agent for preliminary detection of diabetic retinopathy using public datasets. *Intelligence-Based Medicine*, 3:100022, 2020.
31. Prasanna Porwal, Samiksha Pachade, Manesh Kokare, Girish Deshmukh, Jaemin Son, Woong Bae, Lihong Liu, Jianzong Wang, Xinhui Liu, Liangxin Gao, et al. Idrid: Diabetic retinopathy–segmentation and grading challenge. *Medical image analysis*, 59:101561, 2020.
32. Hui Wang, Guohui Yuan, Xuegong Zhao, Lingbing Peng, Zhuoran Wang, Yanmin He, Chao Qu, and Zhenming Peng. Hard exudate detection based on deep model learned information and multi-feature joint representation for diabetic retinopathy screening. *Computer methods and programs in biomedicine*, 191:105398, 2020.
33. Monika Khandelwal and Ranjeet Kumar Rout. Deepprms: advanced deep learning model to predict protein arginine methylation sites. *Briefings in Functional Genomics*, page elae001, 2024.
34. Emran Saleh, Jerzy Błaszczyński, Antonio Moreno, Aida Valls, Pedro Romero-Aroca, Sofia de la Riva-Fernandez, and Roman Słowiński. Learning ensemble classifiers for diabetic retinopathy assessment. *Artificial intelligence in medicine*, 85:50–63, 2018.
35. Nitigya Sambyal, Poonam Saini, Rupali Syal, and Varun Gupta. Modified u-net architecture for semantic segmentation of diabetic retinopathy images. *Biocybernetics and Biomedical Engineering*, 40(3):1094–1109, 2020.
36. Qi Sheng You, Jie Wang, Yukun Guo, Shaohua Pi, Christina J Flaxel, Steven T Bailey, David Huang, Yali Jia, and Thomas S Hwang. Optical coherence tomography angiography avascular area association with 1-year treatment requirement and disease progression in diabetic retinopathy. *American journal of ophthalmology*, 217:268–277, 2020.
37. Srinivas R Sadda, Muneeswar G Nittala, Wongsiri Taweebanjongsin, Aditya Verma, Swetha B Velaga, Ahmed Roshdy Alagorie, Connie M Sears, Paolo S Silva, and Lloyd P Aiello. Quantitative assessment of the severity of diabetic retinopathy. *American Journal of Ophthalmology*, 218:342–352, 2020.
38. Yi-Peng Liu, Zhanqing Li, Cong Xu, Jing Li, and Ronghua Liang. Referable diabetic retinopathy identification from eye fundus images with weighted path for convolutional neural network. *Artificial intelligence in medicine*, 99:101694, 2019.
39. T Jemima Jebaseeli, C Anand Deva Durai, and J Dinesh Peter. Retinal blood vessel segmentation from diabetic retinopathy images using tandem pcnn model and deep learning based svm. *Optik*, 199:163328, 2019.
40. Stela Vujosevic, Stephen J Aldington, Paolo Silva, Cristina Hernández, Peter Scanlon, Tunde Peto, and Rafael Simó. Screening for diabetic retinopathy: new perspectives and challenges. *The Lancet Diabetes & Endocrinology*, 8(4):337–347, 2020.

7 Innovative Approaches to Bilirubin Detection: Using Smart Sensor Technologies for Enhanced Diagnostic Capabilities

Kabita Kumari, Sharvan Kumar Pahuja, and Sanjeev Kumar

7.1 INTRODUCTION

Bilirubin, a pigment known for its distinctive golden colour, plays a crucial function in the complex mechanisms of the human body. Bilirubin is an important indicator of several physiological processes and is produced when heme, a component of haemoglobin in red blood cells, breaks down. The significance of jaundice extends beyond its visual appearance. It is a crucial indicator of liver health and the complex interaction between blood components. The identification of bilirubin in the bloodstream by the German physician Rudolf Virchow in 1847 was a crucial breakthrough in comprehending the composition and the pathophysiology of blood. Virchow's meticulous blood examination identified the yellow pigment bilirubin as a notable constituent [1]. Bilirubin deposition in tissues occurs exclusively in the presence of elevated serum bilirubin levels, indicating either liver disease or, less frequently, a disorder in bilirubin metabolism. The extent of serum bilirubin elevation can be approximated through physical examination. Slight increases in serum bilirubin are best observed by examining the sclerae of the eyes. Sclerae, rich in elastin, have a specific affinity for bilirubin, and the presence of scleral icterus signifies a serum bilirubin level of at least 51μmol/L (3mg/dl). Detecting scleral icterus becomes more challenging in rooms illuminated with fluorescent lighting. Another reliable marker for elevated serum bilirubin is the darkening of urine, resulting from the renal excretion of conjugated bilirubin. Patients commonly characterize their

DOI: 10.1201/9781032660714-7

urine as resembling tea or cola. The presence of bilirubin signifies an increase in the direct serum bilirubin fraction, indicating the existence of liver or biliary disease.

The estimated bilirubin production in healthy adults is 3.5-4.0 mg/kg body weight per day. However, it is twice as 8.5 ±2.3 mg. kg of body weight in newborn infants per day [2]. In human serum, there are two forms of bilirubin: unconjugated (Bu) and conjugated (Bc) [3]. When conjugated bilirubin combines with glucuronic acid, it forms a water-soluble compound, while unconjugated bilirubin binds with albumin, rendering it water-soluble [4]. A minimal amount of unbound, free bilirubin not associated with albumin is a crucial indicator of bilirubin poisoning. Typically, the levels of direct bilirubin in human blood fall within the range of (0.06–0.3) mg/dL, and the total bilirubin concentration is approximately 1.23 mg/dL. Higher bilirubin levels may be a sign of jaundice, hepatitis, cirrhosis, blood infections, transfusion reactions or haemolytic diseases in newborns that lead to cell destruction, Gilbert's syndrome, Crigler-Najjar syndrome, and other health problems that could be very dangerous. On the other hand, low bilirubin levels are connected to conditions like anaemia and coronary artery diseases [5]. In neonates, untreated high concentrations of unconjugated free bilirubin can result in serious consequences such as brain damage, including hearing loss and the potentially fatal syndrome known as "Kernicterus" [6]. Checking bilirubin levels and other liver enzyme tests can help determine if high bilirubin is caused by hemolysis or another liver disease that isn't hemolysis. Pathological levels of bilirubin often coincide with the accumulation of its oxidized form, biliverdin, further emphasizing the diagnostic significance of bilirubin. Given this importance, developing an affordable detection device for bilirubin becomes a matter of great interest as bilirubin is a valuable measure that can help with diagnosis and prognosis [7].

7.1.1 CHEMICAL STRUCTURE OF BILIRUBIN

The natural degradation of heme occurs through a sequence of intermediate stages, with the pivotal component being the orange-yellow bile pigment known as 'bilirubin.' This chromophore is accountable for the colouration observed in different types of jaundice. In young humans, erythrocytes last for about three months, and their destruction causes bilirubin production of about 300mg each day. However, there were some early doubts arising (for cyclisation between the C18 vinyl group and the Oxygen function at C19) for the gross chemical structure of bilirubin has been in reasonable disputes for many years [8]. The chemical structure of bilirubin was elucidated in 1937 by Fischer and Orth, revealing it as a tetrapyrrole closely resembling haemoglobin [9, 10]. Successful synthesis for bilirubin structure was reported in 1942, shown in Figure 7.1(a). Again in 1976, X-ray crystallography demonstrated that the structure took on a bis-lactam form, a finding further supported by 15N NMR spectroscopy studies in the 1980s [8, 11] shown in Figure 7.1b. A simplification of the Bis-Lactum form of the bilirubin is presented again in Figure 7.1(c). The bilirubin structure presented by Fischer and Plieninger in Figure 7.1 lacked details on stereochemistry, leading to inconsistent descriptions over the years, with variations

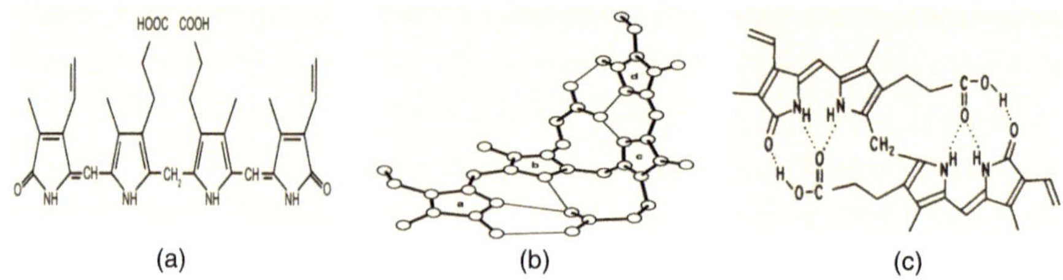

FIGURE 7.1 Structure of Bilirubin. (a) The planar structure of bilirubin was described by Fischer and Plieninger. (b) The bis-lactam structure of bilirubin by Bonnett, Davies, and Hrusthouse, with hydrogen bonds indicated by thin lines. (c) A simplified version of structure (b) [12].

between the correct structure 4Z,15Z and 4E,15E configurations. Subsequent studies have firmly established that bilirubin predominantly exists as the 4Z,15Z isomer, and this configuration may hold significance for the process of entry into the brain and neurotoxicity caused by it [12].

7.1.2 METABOLISM OF BILIRUBIN

The journey of bilirubin metabolism commences with the natural breakdown of red blood cells, liberating heme into the bloodstream. Figure 7.2 delineates the metabolic pathway of bilirubin. Heme, predominantly sourced from ageing red blood cells, contributes to about 20% of daily bilirubin production, while proteins

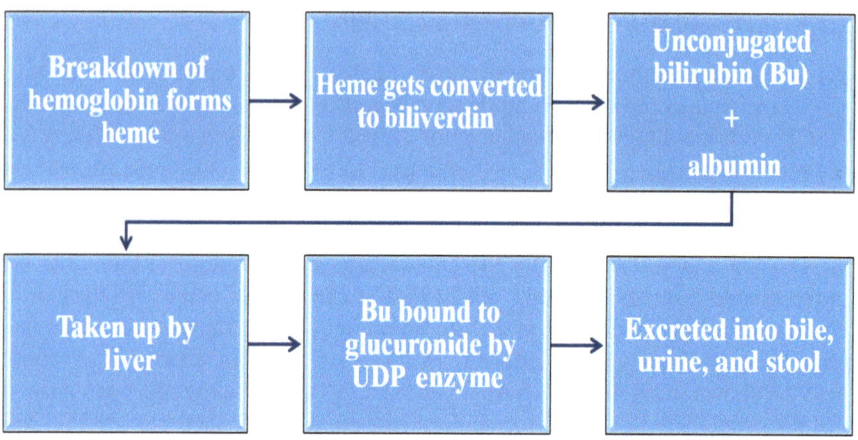

FIGURE 7.2 Block representation for metabolic pathway of bilirubin formation and excretion from body.

like cytochrome P450 isoenzymes and myoglobin also play a role. Monocytic macrophages in the spleen and bone marrow, as well as hepatic Kupffer cells, synthesize heme [13]. Through hemo-oxygenase, heme is transformed into Biliverdin, a linear molecule comprising four pyrrole rings, with the concomitant generation of iron and carbon monoxide [14]. Biliverdin reductase converts Biliverdin into unconjugated bilirubin. To prevent isomerization and facilitate safe transportation throughout the body, bilirubin binds to albumin with a dissociation constant (Kd) of approximately 10^7–10^8 mol/L. Via sinusoids, albumin-bound bilirubin enters the liver. Inside liver cells, bilirubin binds to water-soluble proteins known as ligandins or Y proteins, inhibiting the excretion of absorbed bilirubin. In the smooth endoplasmic reticulum, bilirubin undergoes conjugation with glucuronic acid by UDPGT-1A1, yielding bilirubin glucuronides [15]. Figure 7.3 illustrates the comprehensive bilirubin production and excretion phases. Conjugated bilirubins return to the cytosol and can be excreted into bile or plasma through the canalicular or sinusoidal membrane, respectively, facilitated by MRP2 (multidrug-related protein-2) [16]. The energy-intensive process of excreting conjugated bilirubin results in bile being significantly more concentrated than hepatocyte cytoplasm. Due to its water solubility, conjugated bilirubin is less prone to reabsorption by the gut, aided by microbial and intestinal mucosa β-glucuronidase. Unconjugated bilirubin may revert to its unconjugated form through hydrolysis, facilitated by β-glucuronidase. The intestinal mucosa can reabsorb this

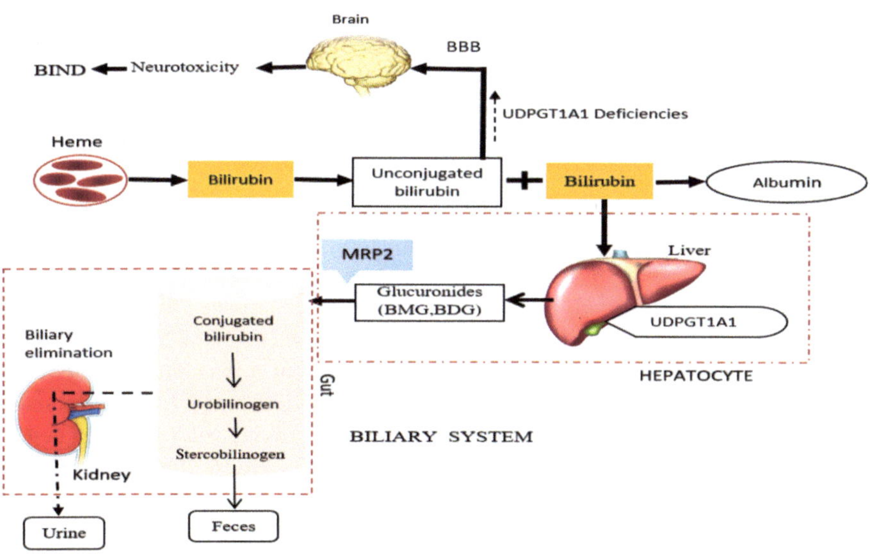

FIGURE 7.3 Schematic representation of metabolic process of bilirubin [15].

unconjugated bilirubin, re-entering the enterohepatic circulation. Roughly 25% of excreted bilirubin is reabsorbed via the bile duct [17].

Within the colon, anaerobic gut bacteria convert the remaining unconjugated bilirubin into colourless tetrapyrroles, such as stercobilinogen, urobilinogen, or urobilinogen, which are collectively referred to as urobilinogen. Approximately 20% of the urobilinogen generated each day is absorbed back from the gut, contributing to enterohepatic recirculation. The liver absorbs most of this urobilinogen via the portal vein and then is eliminated back into bile. Approximately 2-5% of the substance reaches the bloodstream and is then excreted by the kidneys, making it detectable in urine.

7.1.3 WHY CLINICAL ESTIMATION OF BILIRUBIN IS NEEDED?

Clinical estimation of bilirubin is important for several reasons in the medical field, particularly in the assessment of liver and haematological disorders. Conditions that disrupt the normal bilirubin metabolism can lead to elevated concentrations of this pigment in the blood vessels. When the concentration of bilirubin in the blood exceeds 1 mg/dL, it can result in the deposition of bilirubin in various tissues, especially those abundant in elastic fibres like the palate and conjunctiva of eyes [18]. A significant accumulation, typically surpassing 2.5 mg/dL, presents a yellowish discolouration of the skin and mucous membranes, commonly known as 'jaundice'. Jaundice is a clinical sign of hyperbilirubinemia. Elevated levels of this pigment in the bloodstream can lead to its buildup in the brain as it crosses the blood-brain barrier (BBB). This condition, kernicterus, characterized by a yellow-stained nucleus, can potentially cause brain damage. The impairment of brain function occurs through an unknown mechanism and may result in cerebral palsy, and if left untreated, it can be fatal. An elevated level of bilirubin has affected adults as well as neonates. It mostly became vital in neonates due to the improper development of the neonate's liver. Neonatal jaundice has a notable impact on a significant portion of newborns, affecting around 60% of babies born at full term and 80% of those born prematurely [19, 20].

Approximately 140 million newborns worldwide are affected with jaundice during the first two weeks of birth due to its high incidence [21]. Bhutani et al. [19], as illustrated in Figure 7.4(a), were the first to present a mathematical modelling framework in pioneering the evaluation of the worldwide impact of severe jaundice was conducted. The model's estimation suggests that around 18% (approximately 24 million) of the 134 million live births in 2010 were affected by clinically significant jaundice. Furthermore, 0.481 million late-preterm and term neonates experienced severe hyperbilirubinemia (TSB>25 mg/dl), leading to 0.114 million deaths and over 0.063 million survivors dealing with lasting neurological issues ranging from moderate to chronic. Figure 7.4(b) represents the global estimation of impairment caused by kernicterus again in the year 2013, i.e., out of the 480,700 live births vulnerable to Rh disease, 24% faced the risk of neonatal death (estimated at 114,100 with

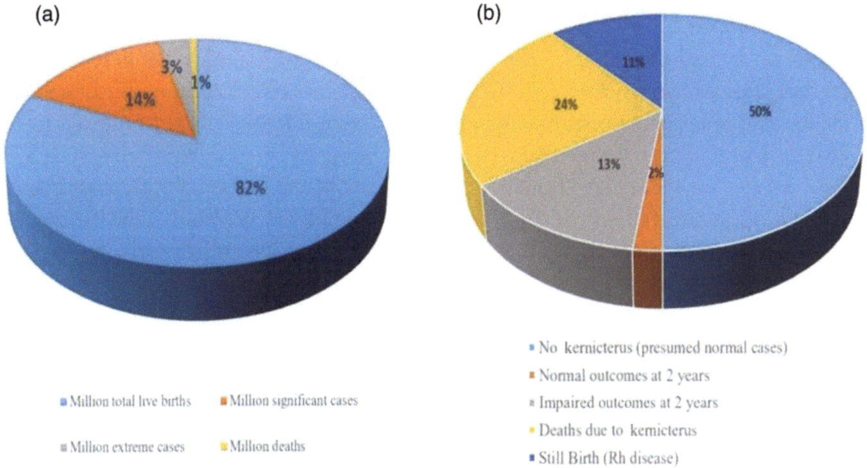

FIGURE 7.4 (a) Global occurrence of substantial jaundice in neonates. (b) Global consequences of persistent impairment resulting from kernicterus.

an uncertainty range of 59,700 to 172,000), 13% were susceptible to kernicterus (amounting to 75,400 with an uncertainty range of 43,200 to 95,800), and 11% were at risk of stillbirths [19]. It's essential to adhere to the guidelines set by the American Academy of Pediatrics (AAP) and the UK National Institute for Health and Care Excellence (NICE), which emphasize the importance of closely monitoring bilirubin levels in newborns and applying suitable treatments as needed [22–24]. Neonatal jaundice poses a significant public health challenge in India, given the high birth rate and the insufficiently developed healthcare infrastructure. To address this issue, the Indian government has launched various initiatives like the National Rural Health Mission (NRHM), aimed at training healthcare personnel and equipping primary health centres with phototherapy devices.

A scheme by the government, 'Janani Shishu Suraksha Karyakram' (JSSK), complements these efforts by providing complimentary care and treatment for newborns with jaundice, which has proven effective in diminishing the prevalence of neonatal jaundice. However, continuous research and development are essential. The focus should be on developing more accurate and economical diagnostic instruments to facilitate early detection and intervention. In summary, clinical estimation of bilirubin is a valuable tool in diagnosing and monitoring various medical conditions, especially those related to liver function and haematology. It furnishes vital data for healthcare practitioners to make well-informed judgements regarding patient care and treatment.

7.2 DIFFERENT APPROACHES FOR THE DETECTION OF BILIRUBIN

Several analytical methods are presently accessible for quantifying bilirubin and its metabolites in serum, urine, and faeces. Various invasive and non-invasive methods are available for detecting jaundice. The measurement of Total Serum Bilirubin (TSB) is a prevalent invasive technique for assessing hyperbilirubinemia levels. This method relies on spectrophotometry and involves obtaining a small blood sample using a syringe and needle [25]. This method is considered the gold standard method for bilirubin estimation. A general invasive process being followed in the clinical laboratories is shown in Figure 7.5. Nonetheless, the assessment of TSB is limited to skilled medical professionals in well-equipped facilities. Additionally, given its invasive nature, this procedure induces stress for both infants and their parents. Since the initial description of bilirubin was in the late 19th century, the predominant techniques for measuring serum bilirubin rely on the bilirubin's reaction with diazotized sulfanilic acid, commonly referred to as the diazo reaction [26].

Numerous bilirubin tests in clinical chemistry laboratories employ diazotization and oxidase methodologies [27]. Under highly acidic conditions with a pH ranging between 1.7 and 2.0, bilirubin and diazo ions undergo a colourimetric reaction, forming a red azo compound through an automated process. The fluorescence intensity of this diazo compound is then measured at 600nm using a photometer to quantify bilirubin levels. While the diazo test for bilirubin is cost-effective and automated, it is vulnerable to hemolysis. It can be affected by lipids and albumin interference, potentially resulting in pH fluctuations and unreliable bilirubin measurements [28, 29].

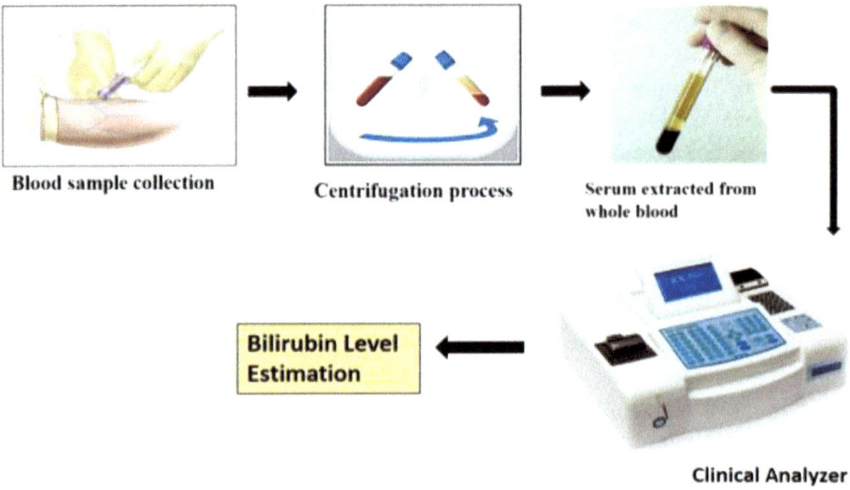

FIGURE 7.5 Laboratory estimation of bilirubin value measurement (invasive method).

The existing bilirubin diazo detection method is only suitable for qualitative analysis because it cannot meet the high accuracy standards needed for quantitative analysis [30, 31]. Bilirubin oxidase (BOD) is categorized as a multi-copper oxidase within the oxidoreductase class. Utilizing metal ions, it facilitates the oxidation of bilirubin into biliverdin, making it a commonly employed reagent in bilirubin testing [32]. Another technique involves chemiluminescence, where bilirubin undergoes redox reactions in aqueous solutions [33]. This process involves the reaction between N-bromosuccinimide and sodium hypochlorite as oxidants, producing intense radiation. A batch chemiluminometer was utilized to analyze these chemiluminogenic reactions. Mechanized analysis of the reactivity between N-bromosuccinimide, hypochlorite, and bilirubin was performed using flow injection and continuous flow analyzers. This method demonstrated sensitivity, with a broader linearity range (0.2-20 μg/mL) and a lower detection limit (50 ng/mL). However, these processes are invasive, and this approach to measure serum bilirubin concentrations (TSB) entails a painful process of blood sampling [34], a procedure that carries various potential long-term repercussions such as the rare occurrence of infection and osteomyelitis [35, 36]. Healthcare workers are also exposed to the risk of needle stick injury with blood sampling. So to overcome these problems, an alternative way, i.e. non-invasive methods, can be employed, or further advancement can be made.

7.3 NON-INVASIVE APPROACHES FOR BILIRUBIN DETECTION

Various developments have been made, and some are ongoing, for the noninvasive or contactless assessment of bilirubin. This section discusses the existing noninvasive modalities and methodologies employed as point–of–care (POC) screening and testing devices for the quantification of bilirubin as well as diseases (e.g., jaundice) caused by bilirubin levels.

7.3.1 VISUAL INSPECTION

Visual assessment is a common method for estimating bilirubin levels, involving observing skin and mucous membrane colouration. The yellow discolouration, known as jaundice, indicates elevated blood bilirubin and is a straightforward, non-invasive means of identifying hyperbilirubinemia in newborns. One assessment method, the Kramer scale, evaluates jaundice severity based on skin appearance, following the cephalocaudal progression principle. This principle notes that newborns' jaundice typically starts on the face and spreads to the chest, abdomen, and extremities as total serum bilirubin (TSB) levels rise [37]. In a study by Purcell and Beeby, researchers investigated whether the progression of jaundice from head to feet in newborns correlated with variations in skin temperature and perfusion across specific anatomical areas: forehead, sternum, lower abdomen, mid-thigh, and sole. The study found that newborns often exhibit increased blood flow to the head and adjacent regions in the early days of life, leading to elevated temperatures and bilirubin accumulation in these areas [38]. However, accurately assessing jaundice severity in newborns receiving outpatient care presents challenges. Some newborns may have significant hyperbilirubinemia without obvious clinical jaundice manifestations. Relying solely on the head-to-toe progression may delay severe jaundice identification and management, risking neurological complications like kernicterus. Additionally, this approach overlooks ethnic and racial differences, potentially exacerbating severity among certain patient populations [39–41].

7.3.2 TRANSCUTANEOUS BILIRUBINOMETRY

Laboratory-based testing for total serum bilirubin (TSB) is the preferred method. However, it might be difficult in resource-limited areas owing to high costs and a lack of trained staff. Non-invasive transcutaneous bilirubin (TcB) devices have become a realistic option for screening in some situations, eliminating the need for eye inspection. These devices use visible light technology to determine serum bilirubin levels by analyzing the skin, providing a near approximation of blood bilirubin content. The forehead and sternum are often used places for testing. Although TcB devices come in various designs, they share a common feature of analyzing skin reflectance spectra, which involves examining the light absorbed and reflected by the skin and subcutaneous tissue when exposed to different wavelengths emitted by the device. The composition of cutaneous chromophores, such as melanin, collagen, oxygenated and reduced haemoglobin, and notably bilirubin, influences the spectra of the reflected light [42, 43]. Table 7.2 (*Continued*) is a compilation of current studies focusing on six distinct types of non-invasive TcB devices. Table 7.2 (*Continued*) includes information such as the research title, the year of publication, the instrument name, the technique or method used, the place or site of measurement, and the study's conclusion.

Kuo et al. [44] pioneered a non-invasive method to detect bilirubin levels by spectrally analyzing colored images using a mini-LED light source. These images were captured with a smartphone and analyzed for the intensity of red (R), green (G), and blue (B) hues. The study aimed to explore the direct correlation between variations in direct bilirubin levels and the spectral characteristics of the test paper image. Notably, the green channel exhibited the highest coefficient of determination 'R' at 0.9313 and the lowest detection limit at 0.56 mg/dL, within the direct bilirubin concentration range of 0.1–2.0 mg/dL. The "AJONEO" device is a novel development in TcB metres. It is designed to take measurements from the newborn nailbed. An observational study was done in Kolkata, India, over a period of 15 months. The study included 2,092 newborn babies with gestational periods between 28 and 40 weeks. The study identified the effective population size as 1,968 newborns. The study found a strong positive linear connection between AJONEO and TSB bilirubin concentrations, with no significant variations seen between right and left nail bed measures [36]. Another non-invasive device that came into the picture was Bilicheck. The BiliChek™ was created by SpectRx Inc (Norcross, GA) and is today widely utilized and promoted by Philips Electronics (Amsterdam, The Netherlands). The BiliChek™ equipment requires a one-time-use tip that needs to be connected to the probe before doing a TcB measurement. This increases the overall cost of operating the instrument. A cohort study was conducted on 58 full-term babies who did not have hemolytic illness. Bilicheck was used to perform transcutaneous measurements on the forehead

TABLE 7.1
Recent Studies in the Development of Non-Invasive Measurement Methods

SNo.	Title	Year	Sample Size	Name of Device	Measurement Technique	Site of Measure-ment	Conclusion
1	Detecting direct bilirubin noninvasively by analysing colour pictures spectrally using a Mini-LED light source.	2022	12 direct bilirubin test papers	NA	Spectum analysis of colored images	NA	Quantifies direct above 1.86 mg/dL bilirubin rapidly and noninvasively
2	Impact of assessment location on TcB measurement	2013	58 term newborns	Bilicheck (SpectRx Inc. Norcross, Georgia, USA)	Transcu-taneous	Forehead and sternum	Accurate readings at sternum than forehead
3	Computer vision for non-invasive early jaundice diagnosis	2023	225 images depicting adult and infant eyes collected from various medical sources	NA	Clustering, filtration and novel metrics based on color space	Sclera	Accuracy of 88.57% is obtained

(Continued)

TABLE 7.2 (*Continued*)
Recent Studies in the Development of Non-Invasive Measurement Methods

SNo.	Title	Year	Sample Size	Name of Device	Measurement Technique	Site of Measurement	Conclusion
4	Evaluating the Bilicare TcB's accuracy as a screening tool for healthy term and late preterm neonates before discharge.	2018	2014 paired samples	Bilicare	Transcutaneous	NA	TSB readings are overestimated at 1 2 mg/dl and underestimated at higher TSB levels
5	Assessment of novel transcutaneous bilirubinometers in neonates	2022	141-full-term-neonates	JAISY v/s JM 105	Transcutaneous	Forehead and sternum	Precise in measuring low to moderate levels of bilirubin
6	Developing a mobile phone camera-based technology for l ow-resource environments.	2022	37 images	Mobile images	Transcutaneous	-	Exhibits promise but need enhancement and further research

and sternum. The correlation and concordance between both techniques and plasma bilirubin were computed. A robust positive connection was seen between the measurements of serum bilirubin at the forehead and sternum with correlation value (r) as 0.704 and p¡0.01 and r value as 0.653 with p¡0.01, respectively [45].

Sreedha et al. [46] proposed a non-invasive method for diagnosing jaundice (elevated bilirubin levels). The proposed method can effectively detect and precisely estimate the extent of illness severity by analyzing the level of yellowness in the sclera. The weighted average of normal (healthy) and jaundiced (unhealthy) eyes at different levels of the disease is calculated by analyzing several eye pictures. The efficacy of the suggested method is assessed by conducting tests on a dataset consisting of 35 eye pictures, and the results are then validated. Among the 35 test photos, the suggested approach accurately assessed the severity of jaundice in 31 photographs. The technique was unsuccessful in the remaining four photos due to an inadequately created mask using the semantic segmentation model. The proposed approach has a testing data accuracy of 88.57%. Peerathat et al. [47] presented a study on the effectiveness and the accuracy of Bilicare transcutaneous meter in late pre-term and term neonates was determined. A cross-sectional study was conducted, and the inclusion criteria consisted of healthy neonates who were delivered at a gestational age (GA) of 35 weeks or later. The blood sample for TSB measurement was collected simultaneously with the normal metabolic test done within the 48-72 hour period before discharge. The association between TSB levels varied between 2.53 and 16.69 mg/dL, with an average (standard deviation) of 9.79 (2.83) mg/dL. The TcB levels ranged from 3.10 to 16.60 mg/dL, with a mean of 10.49 i.e (2.47) mg/dL.

An evaluation of the performance of a novel transcutaneous bilirubinometer called JAISY was presented by Norman et al. [48]. A total of 930 bilirubin measures were conducted on 141 newborn children who were delivered near-term or at term and were between 1 and 6 days old. Using an earlier developed and validated device, JM 105, the obtained results of these measurements were then compared. The measured bilirubin levels ranged from (0–18.8) mg/dl. There was a strong association between bilirubin measurements obtained from the two devices on the forehead, i.e., the correlation coefficient (r) is 0.94, and the chest r value is 0.94. Harrison-Smith et al. [49] presented a mobile phone-based TcB technique and validated it with a 37-patient multi-ethnic pilot trial. An adapter for filtered red-green-blue (RGB) mobile phone camera reflectance measurements was designed using Monte Carlo simulations of neonatal skin reflectance. This showed a link between total serum bilirubin and mobile-phone TcB estimated bilirubin with R2= 0.42 and Bland-Altman limits of agreement of +6.4 to -7.0 mg/dL. Various research has been done and is ongoing in developing non-invasive techniques as point-of-care devices for bilirubin estimation.

The adoption of TcB has grown prevalent in childbirth centres and outpatient departments due to its prompt, easy, non-intrusive and painless characteristics. However, it is crucial to acknowledge that TcB determination is a screening test, and values nearing therapeutic thresholds (e.g., initiation of phototherapy) should be verified through TSB measurements. While having various applications and advantages, Tcb devices also possess some drawbacks in terms of skin pigmentation and race; ambient light may affect the output, effect of phototherapy on TcB reading, etc. [50] [51]. Further research needs to be carried out to establish the long-term benefits and applications of TcB screening. Nonetheless, documented positive short-term outcomes provide a foundation for ongoing investigation.

7.4 CONCULSION

In summary, jaundice is a common issue in neonatal care, often appearing in the first week of life and leading to frequent early hospital readmissions. Current standard practices for assessing bilirubin levels involve invasive TSB measurements, considered the gold standard, and non-invasive TcB assessments using specialized meters. While TSB is direct and invasive, TcB aims to estimate jaundice through optical signals on blanched skin, reducing the need for invasive procedures and minimizing distress for newborns and parents. Non-invasive approaches should complement assessment rather than replace treatment. Future studies on TcB should give primary importance to enhancing the accuracy of devices while considering factors such as skin pigmentation, prenatal age, and postnatal age. Consider utilizing innovative methods such as multispectral imaging and fluorescence spectroscopy to estimate infant bilirubin levels. With advancements in electronic devices, there is potential for a more reliable and safe method for measuring serum bilirubin in newborns. The growing trend of health sensing through smartphones opens the possibility of a non-invasive device using smartphones as a cost-effective and accurate alternative for managing neonatal jaundice in clinical settings.

REFERENCES

1. Natalie Guirguis, Antoine-Xavier Bertrand, Christopher F. Rose, and Simon Matoori. 175 Years of Bilirubin Testing: Ready for Point-of-Care? *Advanced Healthcare Materials*, 12(18):1–8, jul 2023.
2. M. Jeffrey Maisels, Ambadas Pathak, Nicholas M. Nelson, David G. Nathan, and Clement A. Smith. Endogenous production of carbon monoxide in normal and erythroblastotic newborn infants. *Journal of Clinical Investigation*, 50(1):1–8, jan 1971.
3. Johan Fevery. Bilirubin in clinical practice: a review. *Liver International*, 28(5):592–605, may 2008.
4. Irene Taurino, Viviane Van Hoof, Giovanni De Micheli, and Sandro Carrara. Superior sensing performance of multi-walled carbon nanotube-based electrodes to detect unconjugated bilirubin. *Thin Solid Films*, 548:546–550, dec 2013.
5. Sheilą Sherlock and James Dooley, editors. *Diseases of the Liver and Biliary System.* Wiley, dec 2001.
6. Charles E. Ahlfors and Anne E. Parker. Evaluation of a Model for Brain Bilirubin Uptake in Jaundiced Newborns. *Pediatric Research*, 58(6):1175–1179, dec 2005.
7. Brittany AuYoung, Akshay Gutha Ravichandran, Divykumar Patel, Nisarg Dave, Achal Shah, Brianna Wronko-Stevens, Franklin Bettencourt, Reshma Rajan, and Nidhi Menon. A Low-Cost Paper-Based Device for the Colorimetric Quantification of Bilirubin in Serum Using Smartphone Technology. *Frontiers in Chemistry*, 10(July):1–11, jul 2022.
8. Raymond Bonnett, John E Davies, and Michael B Hursthouse. Structure of bilirubin. *Nature*, 262(5566):326–328, jul 1976.

9. Hans Fischer and Hans Plieninger. Synthese des Biliverdins (Uteroverdins) und Bilirubins, der Biliverdine XIII α und III α , sowie der Vinylneoxanthosaure. *Hoppe-Seyler ́s Zeitschrift fur physiologische Chemie*, 274(1-6):231–260, jan 1942.
10. Monika Khandelwal, Sabha Sheikh, Ranjeet Kumar Rout, Saiyed Umer, Saurav Mallik, and Zhongming Zhao. Unsupervised learning for feature representation using spatial distribution of amino acids in aldehyde dehydrogenase (aldh2) protein sequences. *Mathematics*, 10(13):2228, 2022.
11. Poul E. Hansen and Hans J. Jakobsen. A natural abundance 15 N NMR investigation of bilirubin IX-α. *Organic Magnetic Resonance*, 22(10):668–670, oct 1984.
12. Thor W. R. Hansen, Ronald J. Wong, and David K. Stevenson. Molecular Physiology and Pathophysiology of Bilirubin Handling by the Blood, Liver, Intestine, and Brain in the Newborn. *Physiological Reviews*, 100(3):1291–1346, jul 2020.
13. J Jacobsen. Binding of bilirubin to human serum albumin - determination of the dissociation constants. *FEBS letters*, 5(2):112–114, oct 1969.
14. Serge Erlinger, Irwin M Arias, and Daniel Dhumeaux. Inherited Disorders of Bilirubin Transport and Conjugation: New Insights Into Molecular Mechanisms and Consequences. *Gastroenterology*, 146(7):1625–1638, jun 2014.
15. Andrew Rowland, John O Miners, and Peter I Mackenzie. The UDP-glucuronosyltransferases: Their role in drug metabolism and detoxification. *The International Journal of Biochemistry & Cell Biology*, 45(6):1121–1132, jun 2013.
16. Kabita Kumari, Shravan Kumar Pahuja, and Sanjeev Kumar. An Evolution of Bilirubin Physiology and Analysis. *Current Signal Transduction Therapy*, 18(2), jul 2023.
17. Armando Raúl Guerra Ruiz, Javier Crespo, Rosa Maria López Martínez, Paula Iruzubieta, Gregori Casals Mercadal, Marta Lalana Garcés, Bernardo Lavin, and Manuel Morales Ruiz. Measurement and clinical usefulness of bilirubin in liver disease. *Advances in laboratory medicine*, 2(3):352–372, aug 2021.
18. Kiyoshi Ichihara, Yesim Ozarda, Julian H Barth, George Klee, Ling Qiu, Rajiv Erasmus, Anwar Borai, Svetlana Evgina, Tester Ashavaid, Dilshad Khan, Laura Schreier, Reynan Rolle, Yoshihisa Shimizu, Shogo Kimura, Reo Kawano, David Armbruster, Kazuo Mori, and Binod K Yadav. A global multicenter study on reference values: 1. Assessment of methods for derivation and comparison of reference intervals. *Clinica chimica acta; international journal of clinical chemistry*, 467:70–82, apr 2017.
19. Vinod K Bhutani, Ann R Stark, Laura C Lazzeroni, Ronald Poland, Glenn R Gourley, Steve Kazmierczak, Linda Meloy, Anthony E Burgos, Judith Y Hall, and David K Stevenson. Predischarge screening for severe neonatal hyperbilirubinemia identifies infants who need phototherapy. *The Journal of pediatrics*, 162(3):477–482.e1, mar 2013.
20. National Institute for Health Excellence and Care. Jaundice in newborn babies under 28 days. *NICE Quality Standard*, (October):1–29, 2014.
21. UNICEF. The State of the World's Children 2017. *Children in a Digital World*, 2017.
22. National Institute for Health and Care Excellence (NICE). Neonatal jaundice. *National Collaborating Centre for Women's and Children's Health*, 2010.
23. Subcommittee on Hyperbilirubinemia. Management of Hyperbilirubinemia in the Newborn Infant 35 or More Weeks of Gestation. *Pediatrics*, 114(1):297–316, 2004.
24. World Bank Group World Health Organization, United Nations Children's Fund. Postnatal Care for Mothers and Newborns: Highlights from the World Health Organization 2013 Guidelines. *World Health Organization*, (April):1–8, 2015.
25. Fatema-Tuz-Zohra Khanam, Ali Al-Naji, Asanka G. Perera, Danyi Wang, and Javaan Chahl. Non-invasive and non-contact automatic jaundice detection of infants based on

random forest. *Computer Methods in Biomechanics and Biomedical Engineering: Imaging & Visualization*, 11(6):2516–2529, nov 2023.
26. Paul Y. Kwo, Stanley M. Cohen, and Joseph K. Lim. ACG Clinical Guideline: Evaluation of Abnormal Liver Chemistries. *American Journal of Gastroenterology*, 112(1):18–35, 2017.
27. Sundaram Ellairaja, Kathiravan Shenbagavalli, Sarkaraisamy Ponmariappan, and Vairathevar Sivasamy Vasantha. A green and facile approach for synthesizing imine to develop optical biosensor for wide range detection of bilirubin in human biofluids. *Biosensors and Bioelectronics*, 91:82–88, 2017.
28. Shib Shankar Singha, Suchanda Mondal, Tara Shankar Bhattacharya, Laboni Das, Kamalika Sen, Biswarup Satpati, Kaustuv Das, and Achintya Singha. Au nanoparticles functionalized 3D-MoS2 nanoflower: An efficient SERS matrix for biomolecule sensing. *Biosensors and Bioelectronics*, 119:10–17, 2018.
29. Neha Dhungana, Cory Morris, and Matthew D. Krasowski. Operational impact of using a vanadate oxidase method for direct bilirubin measurements at an academic medical center clinical laboratory. *Practical Laboratory Medicine*, 8(May):77–85, 2017.
30. M S Devgun and C Richardson. Direct bilirubin in clinical practice — Interpretation and haemolysis interference guidance reassessed. *Clinical Biochemistry*, 49(18):1351–1353, 2016.
31. Steven C Kazmierczak, F Robertson, Paul G Catrou, Kimberly P Briley, Bill Kreamer, and Glenn Gourley. Bilirubin in Newborns : Comparison with HPLC and an Automated Diazo Method. *Clinical Chemistry*, 48(7):1096–1097, 2002.
32. Rachna Rawal, Poonam R Kharangarh, Sudhir Dawra, and Preetam Bhardwaj. Synthesis, characterization and immobilization of bilirubin oxidase nanoparticles (BOxNPs) with enhanced activity: Application for serum bilirubin determination in jaundice patients. *Enzyme and Microbial Technology*, 143:109716, 2021.
33. L P Palilis, A C Calokerinos, and N Grekas. Chemiluminescence arising from the oxidation of bilirubin in aqueous media. *Analytica Chimica Acta*, 333(3):267–275, 1996.
34. S H Yap, I Mohammad, and C A Ryan. Avoiding painful blood sampling in neonates by transcutaneous bilirubinometry. *Irish journal of medical science*, 171(4):188–190, 2002.
35. Selçuk Yuksel, Gulten Yuksel, Selim Oncel, and Elfi Divanli. Osteomyelitis of the calcaneus in the newborn: An ongoing complication of Guthrie test. *European Journal of Pediatrics*, 166(5):503–504, 2007.
36. Animesh Halder, Aniruddha Adhikari, Ria Ghosh, Soumendra Singh, Amrita Banerjee, Nilanjana Ghosh, Arnab Madhab Bhattacharya, Shrabani Mandal, Prantar Chakrabarti, Debasis Bhattacharyya, Hatem M. Altass, Moataz Morad, Saleh A. Ahmed, Asim Kumar Mallick, and Samir Kumar Pal. Large scale validation of a new non-invasive and non-contact bilirubinometer in neonates with risk factors. *Scientific Reports*, 10(1):1–14, 2020.
37. Arieh Riskin, Ada Tamir, Amir Kugelman, Miri Hemo, and David Bader. Is Visual Assessment of Jaundice Reliable as a Screening Tool to Detect Significant Neonatal Hyperbilirubinemia? *The Journal of Pediatrics*, 152(6):782–787.e2, jun 2008.
38. Natalie Purcell and Philip J Beeby. The influence of skin temperature and skin perfusion on the cephalocaudal progression of jaundice in newborns. *Journal of Paediatrics and Child Health*, 45(10):582–586, oct 2009.
39. Sri Devi, Manjubala Dash, and Felicia Chitra. Detection of Neonatal Jaundice among the Newborn Using Kramers Criteria. *Epidemiology: Open Access*, 08(04), 2018.

40. Mahendra Tri Arif Sampurna, Muhammad Pradhika Mapindra, Muhammad Pradhiki Mahindra, Kinanti Ayu Ratnasari, Siti Annisa Dewi Rani, Kartika Darma Handayani, Dina Angelika, Agus Harianto, Martono Tri Utomo, Risa Etika, and Pieter J.J. Sauer. Kramer score, an evidence of its use in accordance with indonesian hyperbilirubinemia published guideline. *International Journal of Environmental Research and Public Health*, 18(11), 2021.
41. Berthe A.M. van der Geest, Malou J.S. de Mol, Ivana S.A. Barendse, Johanna P. de Graaf, Loes C.M. Bertens, Marten J. Poley, Erwin Ista, René F. Kornelisse, Irwin K.M. Reiss, Eric A.P. Steegers, Jasper V. Been, Martin G.A. Baartmans, Jolita Bekhof, Harry Buijs, Jan Erik Bunt, Peter H. Dijk, Christian V. Hulzebos, Ralph W.J. Leunissen, Ben J.P.W. Snoeren, Bente de Vries, and Leo Wewerinke. Assessment, management, and incidence of neonatal jaundice in healthy neonates cared for in primary care: a prospective cohort study. *Scientific Reports*, 12(1):1–12, 2022.
42. William D Engle, Gregory L Jackson, and Nancy G Engle. Transcutaneous bilirubinometry. *Seminars in Perinatology*, 38(7):438–451, 2014.
43. Steven L. Jacques, Iyad S. Saidi, Anne Ladner, and David Oelberg. Developing an optical fiber reflectance spectrometer to monitor bilirubinemia in neonates. In Steven L. Jacques, editor, *Laser-Tissue Interaction VIII*, volume 2975, pages 115–124, jun 1997.
44. Hsin Ching Kuo, Zhi Ting Ye, Shen Fu Tseng, Shang Xuan Tsou, Shih Wei Huang, and Chun Wei Tsai. Noninvasive direct bilirubin detection by spectral analysis of color images using a Mini-LED light source. *Discover Nano*, 18(1), 2023.
45. Cristiane M.aria Conceição, Maria F.ernanda Pellegrino da Silva Dornaus, Maria A.parecida Portella, Alice D.Agostini Deutsch, and Celso M.oura Rebello. Influence of assessment site in measuring transcutaneous bilirubin. *Einstein (São Paulo, Brazil)*, 12(1):11–15, 2014.
46. B Sreedha, Prashant R Nair, and Reevu Maity. Non-invasive early diagnosis of jaundice with computer vision. *Procedia Computer Science*, 218:1321–1334, 2023.
47. Peerathat Chokemungmeepisarn, Watcharee Tantiprabha, Shanika Kosarat, and Satit Manopunya. Accuracy of the Bilicare™ transcutaneous bilirubinometer as the predischarge screening tool for significant hyperbilirubinemia in healthy term and late preterm neonates. *The Journal of Maternal-Fetal & Neonatal Medicine*, 33(1):57–61, jan 2020.
48. Mikael Norman, Huseyin Aytug, and Hasan Basri Celebi. Evaluation of a new transcutaneous bilirubinometer in newborn infants. *Scientific reports*, 12(1):5835, apr 2022.
49. Brandon Harrison-Smith, Alexander P. Dumont, Mohammed Shahriar Arefin, Yu Sun, Nuradeen Lawal, Dorianna Dobson, Amy Nwaba, Sarah Grossarth, Abdulsalam Muhammed Paed, Zubaida L. Farouk, Jorn-Hendrik Weitkamp, and Chetan A. Patil. Development of a mobile phone camera-based transcutaneous bilirubinometer for low-resource settings. *Biomedical Optics Express*, 13(5):2797, 2022.
50. Y Yamauchi and I Yamanouchi. Factors affecting transcutaneous bilirubin measurement: effect of daylight. *Acta paediatrica Japonica : Overseas edition*, 33(5):658–662, oct 1991.
51. Durgesh Kumar and Dinesh Kumar. A Prospective Comparison of Serum and Transcutaneous Bilirubin in Indian Neonates. *Journal of Pediatric Intensive Care*, 11(02):100–104, 2022.

8 Targeted Immunization: Application of Machine Learning in Prediction of IL-4 Inducing Peptides

Sana Farooq, Jewaira Khurshid, and Ishrat Nazeer

8.1 INTRODUCTION

Adaptive immunity or antigen-specific immunity acts as a second line of defence in our body after innate immunity. The adaptive immune system's response to a pathogen involves processing and displaying the pathogen's antigen via Major Histocompatibility Complex (MHC) [1]. Upon recognizing a specific antigen, MHC II binders interact with CD4+T cells, and an antigen-specific cytokine is synthesized by various CD4+T cells accordingly. Based on the cytokines they release, CD4+T can be differentiated into T-cell groups, such as Th1, Th2, Th17, or iTregs [2]. Likewise, MHC II binding peptide sequences can be divided based on the cytokine they induce. In the Th2 cell group, the primary cytokine secreted is Interleukin-4 (IL-4), and the MHC II binder that induces its release is called an IL-4-inducing peptide. IL-4 stimulates the division and differentiation of antigen-presenting cells into B-cells, dendritic cells and macrophages. It also drives the isotype switching of B-cells, allowing them to change the type of immunoglobulin they produce, like IgM, IgG, IgA, and IgE. In recent developments, IL-4 has been extensively used as a modulator in the vaccines of autoimmune diseases like cancer [3], psoriasis [4], multiple myeloma [5], arthritis [6] and various other allergies and infectious diseases. Predicting IL-4-inducing peptides can help in the artificial synthesis of IL-4, eventually developing more effective and targeted immunization strategies. Controlling the production of IL-4 allows for the design of vaccines tailored to specific pathogens, diseases, or patient profiles. Moreover, the artificial synthesis of IL-4 can be explored for its potential in immunotherapy, which can enhance the body's natural immune responses against cancer cells. Figure 8.1 presents the various roles IL-4 takes up in

DOI: 10.1201/9781032660714-8

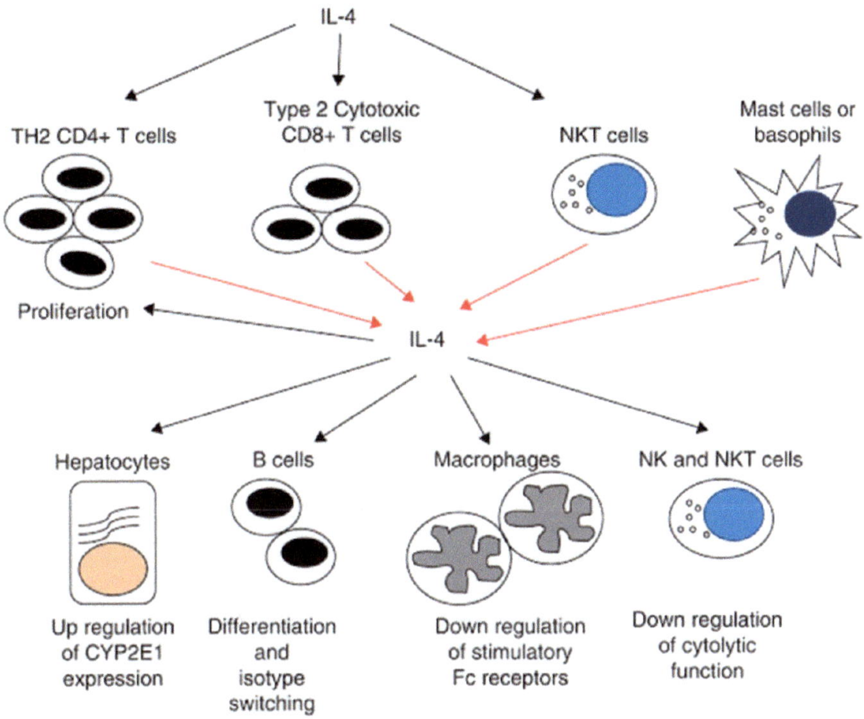

FIGURE 8.1 The multifaceted role of IL-4, within the human immune system, as a key regulatory cytokine [7].

the immune system, highlighting its significance in coordinating the body's immune responses and contributing to immune system regulation [7].

With the growing biological data, traditional methods of predicting protein functions have been replaced with various computational techniques [8]. While these computational techniques provide more consistent results by following pre-defined algorithms and rules, their success largely depends on high-quality training data and expert knowledge. These techniques also lack the flexibility of capturing complex relationships in large biological datasets. However, in the present day, Machine Learning (ML) methods can learn the characteristics of intricate datasets and present them in a more comprehensible manner [9–12]. ML models have not only automated the feature extraction process but also can identify unexpected and novel functions of peptides. Employing ML models enables us to go beyond mere descriptions and offer insights in the shape of testable models. To predict the IL-4-inducing nature of MHC II binding sequences, more than 100 features of these peptide sequences have been extracted for this study. These features provide insight into the compositional and functional properties of peptides, guiding the ML models in predicting their functions. Out of the wide range of peptide features that can be derived from their amino acid sequences, compositional features have been employed to train the predictive model due to the type of information they preserve. This study aims to build a reliable ML system that analyzes these features to distinguish IL-4-inducing peptides from non-inducing ones correctly. The contributions of this study are outlined as follows:

1. To address the challenges of identifying the IL-4 inducing peptides by employing the state-of-art ML techniques.
2. To reduce the dependability of researchers on labor-intensive techniques of identification by building a data-driven predictive model.
3. To perform classification using compositional features of peptide sequences.
4. To increase the use of automatic identification systems and alleviate the computational expenses involved in traditional practices.

8.2 LITERATURE REVIEW

Over the past decade, numerous techniques have been created to predict MHC Class II binding molecules to identify T-helper epitopes that could activate T-helper cells. [13] developed an online tool called the Promiscuous MHC Class-I Binding Peptide Prediction Server (ProPred1) that uses a matrix-based approach to predict the promiscuous region. Another web server, nHLAPred, was created by [14] employing Artificial Neural Network (ANN), Support Vector Machine (SVM), and Quantitative Matrix techniques to predict MHC-binding peptides. Additionally, [15] built PSSMHCpan, a software that relies on Position Specific Scoring Matrices (PSSM) to predict class I peptide-HLA binding affinity. [16] introduced NetMHCpan, a computational model for making quantitative predictions regarding peptide binding to any HLA-A and B-Locus protein with known sequences [17]. In another attempt, [18] developed NetMHC-3.0, a unique web server for predicting MHC class-I affinities for peptides of length 8-11 in human, mouse, and monkey species. For the prediction of MHC class II binding regions in antigenic protein sequences, a graphical web tool called ProPred, based on a matrix prediction algorithm, was engineered by [19]. [20] proposed HLA-DR4Pred, a SVM-based tool for predicting HLA-DRB1*0401 binding peptides in antigen sequences. In addition, [21] built NetMHCII, an alignment algorithm based on ANN for predicting MHC class II binding interactions. Another study by [22] led to the development of HLAncPred to predict non-classical HLA binders.

Recently, researchers have shifted their focus towards developing methods for predicting peptides that can induce specific types of cytokines. Towards this end, [23] proposed CytoPred, a highly accurate method for predicting and categorizing cytokines. In a pioneering step, [24] introduced IL4pred, the first-ever web tool for predicting IL-4-inducing peptides. Multiple attempts have also been made to predict other types of cytokines. [25] conducted a comprehensive study leveraging a wide range of ML techniques to create IL-6Pred, a web server and standalone package for predicting, designing, and screening IL-6-inducing peptides. In a unique study, [26] developed IL-10Pred, which includes a web server, standalone package, and mobile app for designing immunosuppressive peptides based on their potential to induce IL-10. [27] constructed different sequence-based ML models to propose IL17eScan, a tool for identifying peptides that trigger an IL-17 response. Among numerous other tools, [28] introduced IL-13Pred to predict immunoregulatory cytokines, particularly IL-13-inducing peptides. Analysis of proteins as graphs can provide useful insights into the proteins' various functions, folding, and global structural properties.

8.3 MATERIALS AND METHODS

The experimentation was conducted step-wise, and a suitable dataset was procured, followed by its feature extraction. Data has also been preprocessed before the model training to achieve better results. In the same pursuit, model parameters have also undergone optimization. The proposed model has been evaluated using various performance measures for the final step. Figure 8.2 displays the experimental framework of the proposed study, providing a visual representation of the structured methodology and key components employed in this study.

8.3.1 RETRIEVAL OF DATASET

To build a binary classification system, procuring a suitable dataset containing positive and negative peptide sequences (IL-4 inducing and non-inducing) is important. The datasets used in this study have been sourced from publicly available databases, including the Immune Epitope Database (IEDB) [29] and the annotated protein sequence database Swiss-Prot [30]. We specifically collected 985 IL-4-inducing peptides (a positive set) and 1758 non-inducing peptide sequences (a negative set). Each dataset comprises unique sequence IDs and peptide sequences presented in FASTA format. This process yielded a total of 2743 experimentally validated peptide sequences.

8.3.2 FEATURE GENERATION

Training ML models directly on collected peptide sequences provides poor results because of a large volume of data and data redundancy. In this study, peptide sequences have been converted into the following compositional features that provide sequence order-related information and function-related information of the peptides:

1. **Amino acid composition (AAC)**: AAC is the percentage amount of a particular Amino Acid (AA) present in a peptide.
2. **Pseudo Amino Acid Composition (PAAC)**: PAAC represents the relationship between various AAs present in a peptide sequence along with the AA composition.

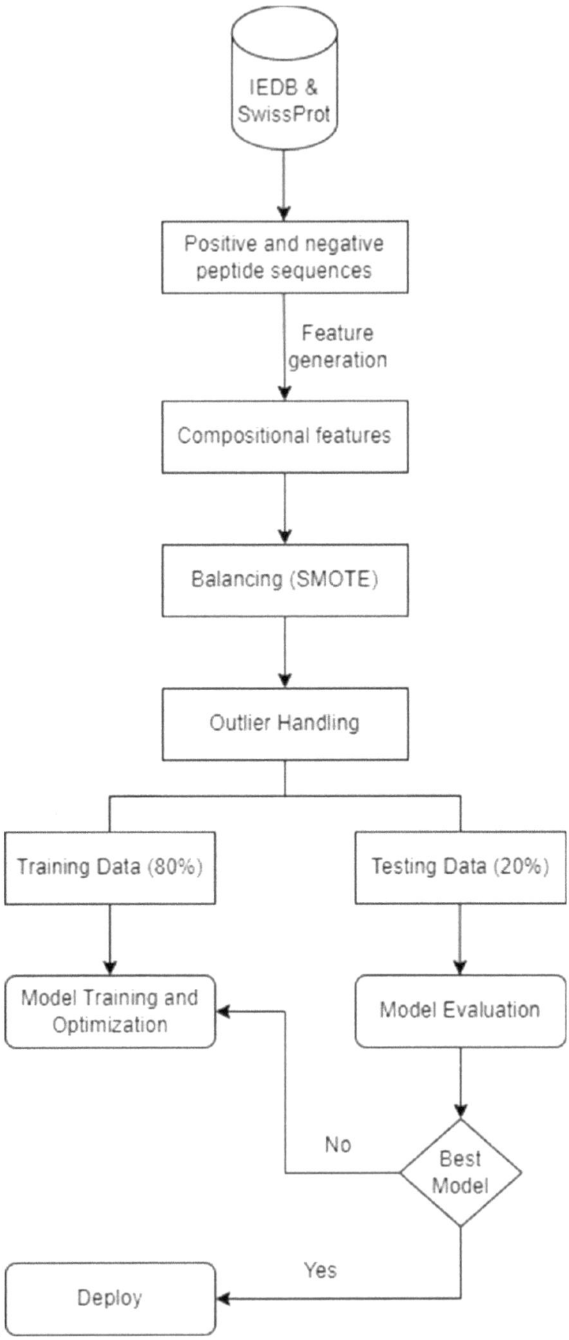

FIGURE 8.2 Overview of the experimental framework, representing the proposed methodology for prediction of IL-4 inducing peptides.

TABLE 8.1
Description of Features in the Dataset

Compositional Feature	Type of Information Preserved	Number of Features
Amino acid composition	Compositional	20
Pseudo amino acid composition	Compositional and sequential	21
Amphiphilic pseudo amino acid composition	Compositional, sequential and aphiphilic	23
Sequence order coupling number	Positional relationships between amino acid pairs	2
Quasi sequence order number	Global sequence information	42

3. **Amphiphilic Pseudo Amino Acid Composition (APAAC)**: APAAC is the measure of autocorrelation of physicochemical properties of AA present in a sequence within that sequence.
4. **Sequence Order Coupling Number (SOCN)**: SOCN refers to the interdependence between AA positions in a peptide sequence.
5. **Quasi Sequence Order Number (QSON)**: QSON determines the quantity and distribution of AA with charged, polar and hydrophobic properties within a peptide sequence.

Following the process of feature generation, the dataset contained a collective of 108 peptide sequence features. The description of features present in the dataset is given in Table 8.1.

8.3.3 DATA PREPROCESSING

As part of the data preprocessing procedures, measures for balancing and handling outliers were applied to transform the data into input for the training model. These two techniques have been explained below:

8.3.3.1 Balancing the dataset

The data employed for this study consists of 985 positive instances and 1758 negative instances, which indicates the imbalanced situation. In the case of binary classification, an imbalanced dataset leads to the biased training of the ML model in favour of a larger class. To avoid this, Synthetic Minority Oversampling Technique (SMOTE) [31] has been deployed to balance the two classes in our dataset. The SMOTE technique balances two classes by oversampling the minority class using linear interpolation to create synthetic samples for that class. The working of SMOTE can described in the following steps:

1. The sampling rate N is determined based on the degree of class imbalance.
2. The k-nearest neighbors of each sample a in the minority class set X are determined by computing the Euclidean distance between a and all other samples in set X.
3. For every sample a, N instances ($a_1, a_2, ..., a_n$) are chosen randomly from its k-nearest neighbors to form the set X_1.
4. A new example is generated for each instance a_k in X1 (where k = 1, 2, 3, ..., N) using Equation 8.1.

$$a' = a + \text{rand}(0,1) \cdot |a - a_k| \quad (8.1)$$

Through the use of SMOTE the minority class has been augmented to match the number of instances in the majority class, resulting in a total of 1758 instances. The distribution of positive and negative classes in the dataset before and after balancing can be represented by Figures 8.3 and 8.4, respectively.

8.3.3.2 Outlier Detection and Handling

Biological datasets are often subjected to rare instances or extreme values, called outliers in the context of ML, which significantly deviate from the expected values.

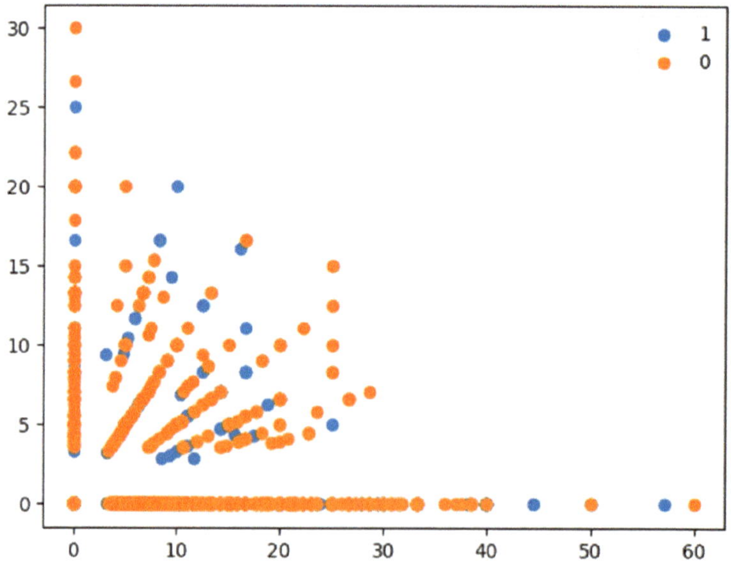

FIGURE 8.3 Distribution of positive and negative classes before balancing.

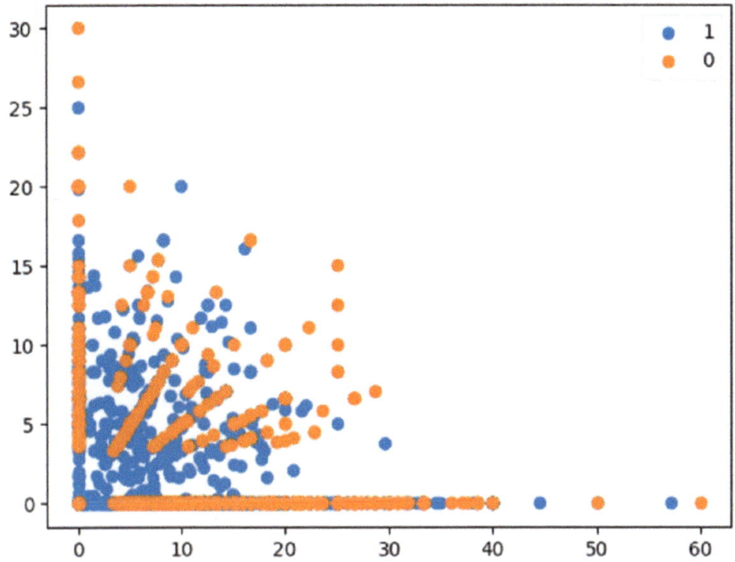

FIGURE 8.4 Distribution of positive and negative classes after balancing.

These can exert a notable impact on the overall performance of the ML model and, therefore, should be addressed. This research identifies outliers using a visualization method, Box plot, managed through a capping technique. A Box plot is a valuable graphical approach for examining data distribution using key statistical measures, including median, lower, and upper quantiles (the 50th, 25th, and 75th percentiles of the dataset, respectively). A box plot draws a box between the upper and lower quantiles, and a line within the box locates the median. Data points falling outside the quantile boundaries are categorized as outliers, with the severity of the deviation determined by whether the inner or outer quantile threshold is breached. The mathematical formulae for the various quantiles are given in Equations 8.2 to 8.5, along with their contribution to outlier detection.

$$\textit{Lower inner Quantile} = Q1 - 1.5 \times \text{IQR} \tag{8.2}$$

Any data point falling below Lower inner Quantile is categorized as a moderate outlier.

$$\textit{Upper inner Quantile} = Q3 + 1.5 \times \text{IQR} \tag{8.3}$$

Any data point exceeding Upper inner Quantile is identified as a moderate outlier.

$$\textit{Lower outer Quantile} = Q1 - 3 \times \text{IQR} \tag{8.4}$$

Data points falling below Lower outer Quantile fall in the category of severe outliers.

$$\textit{Upper outer Quantile} = Q3 + 3 \times \text{IQR} \tag{8.5}$$

Data points exceeding Upper outer Quantile detect as severe outliers.

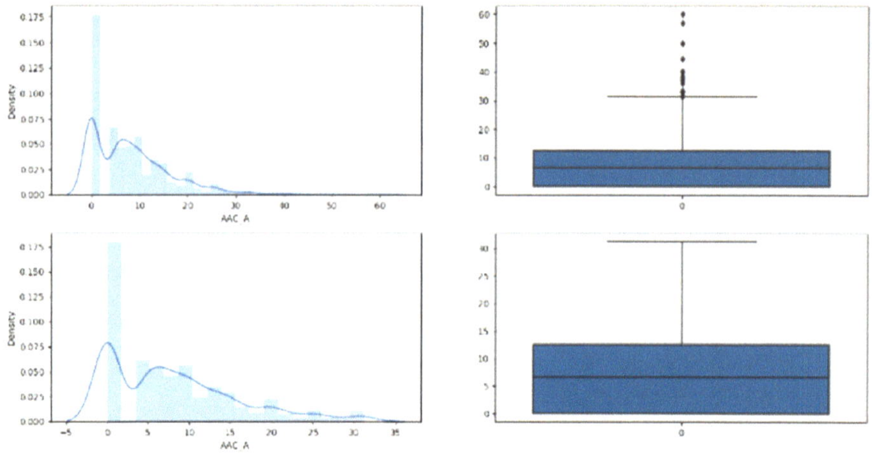

FIGURE 8.5 Outlier handling impact on one of the AAC descriptors (Before and After).

In the above equations, Q1 and Q3 are lower and upper quantiles, i.e., 25th and 75th percentiles of the dataset, respectively. At the same time, IQR (Inter Qunatile Range) refers to the difference between the upper and lower quantile. A practical method for addressing outliers without losing data is the capping technique. In this approach, values exceeding the threshold (quantile) are transformed or replaced with the threshold value. In Figure 8.5, the upper left graph represents the Gaussian distribution of one of the AAC descriptors before the outlier handling process, where deviations from the distribution pattern are identified as outliers. In contrast, the bottom left graph represents the refined distribution post-application of the outlier handling technique. In the same figure, graphs on right side portray the Box plots for the AAC descriptor before and after capping. The box provides a visual representation of the IQR, encapsulating the middle 50% of the data, the whiskers extend from the box to delineate the range of the dataset where individual data points beyond the whiskers (in upper right graph) are identified as outliers. Figures 8.6 to 8.9 illustrate analogous details regarding descriptors found in the other four features of the dataset.

8.3.4 MODEL BUILDING

In the field of ML, a sequence of steps is typically adhered to when creating a predictive model. In the context of this research, three specific steps have been pursued, which are as follows:

1. **Selecting a Suitable ML Model:** Selecting the right ML model is a critical and essential step in the experimental process. It is vital to choose a model that outperforms others and consistently delivers exceptional results. The

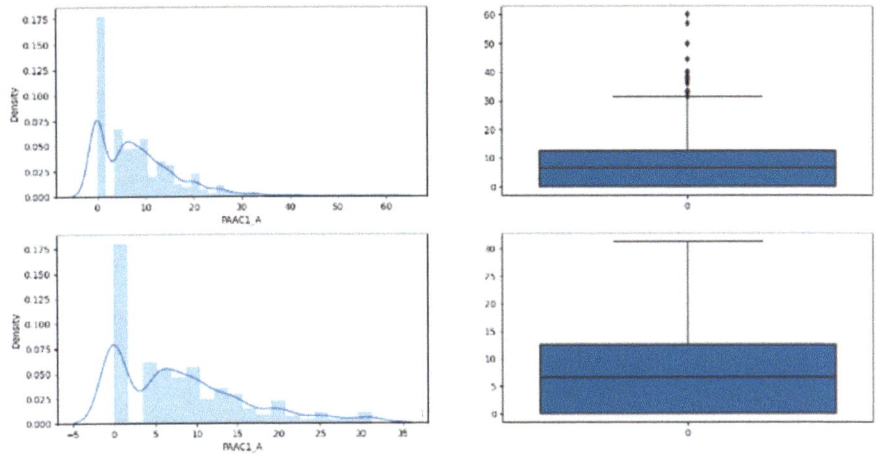

FIGURE 8.6 Outlier handling impact on one of the PAAC descriptors (Before and After).

choice of the most suitable model may vary based on specific factors like the nature of the problem, the dataset used, and the desired outcomes.

2. **Training the ML Model:** The next fundamental step after model selection is training the model. During this phase, the model undergoes a training process and is fine-tuned to attain optimal parameters. The objective is to minimize the disparity between predicted and actual values, striving to create a precise predictor for new data.

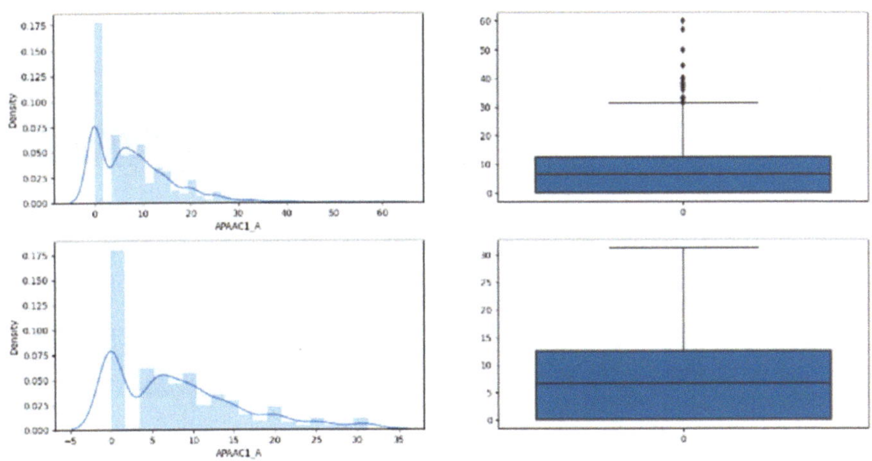

FIGURE 8.7 Outlier handling Impact on one of the APAAC descriptors (Before and After).

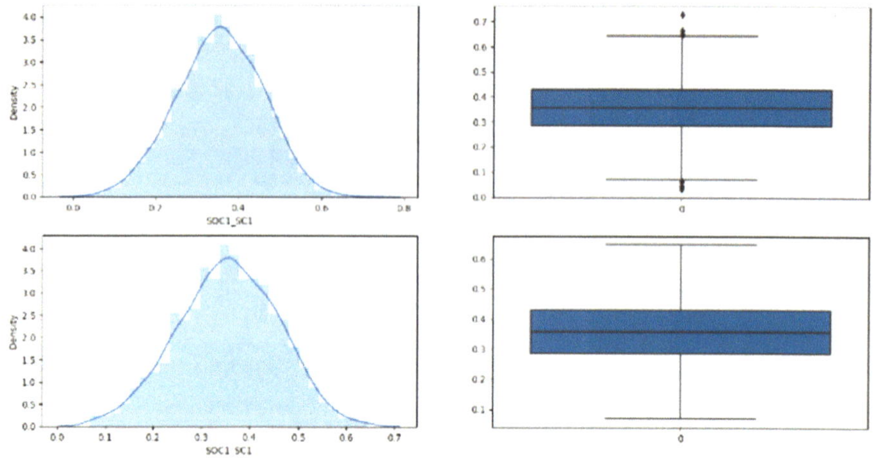

FIGURE 8.8 Outlier handling Impact on one of the SOCN descriptors (Before and After).

3. **Validating the ML Model:** Validating the ML model comes immediately after the training phase and is a crucial step for evaluating the effectiveness of the trained model. This process involves using a designated portion of the dataset, often referred to as the testing set, to assess the model's performance.

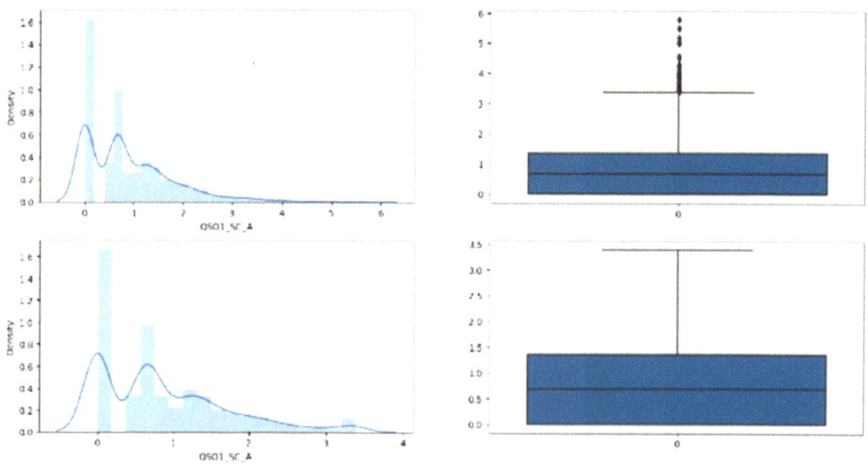

FIGURE 8.9 Outlier handling Impact on one of the QSON descriptors (Before and After).

In this study, we initially trained and tested multiple models, including Logistic Regression and SVM. The initial results depicted a moderate level of accuracy for each of these models. However, the accuracy of tree-based algorithms, i.e. Decision Tree (DT), Random Forest (RF), Extra Tree (ET) classifier and AdaBoost classifier, exhibited a noticeable upward trend. It became apparent that the tree-based model excelled and consistently produced exceptional results. This suggested the potential benefits of utilizing the ensemble models of DT to address the challenges encountered when training a single classifier, such as bias and variance [32]. This approach involves the integration of multiple base learners, where each base learner's contribution combines to achieve the highest level of accuracy and minimize prediction errors. The ensemble of DTs can be achieved through Bagging and Boosting.

Bagging involves sub-sampling the training data through Bootstrapping. Bootstrapping is a process where samples are repeatedly selected with replacements from the dataset, and statistics like mean or median are computed for each sample. This generates a distribution of results for these statistics, facilitating the estimation of standard errors and confidence intervals to obtain less biased estimates. The final prediction is achieved by combining predictions from all learners through majority voting. In this research, RF and ET have been used to implement Bagging. Boosting is a technique that entails using a set of weak learners or base learners to create a robust rule. The progression from developing weak learners into strong learners is accomplished by utilizing the voting or averaging method. The formation of a meta-learner is then achieved through an iterative process, wherein the models are trained so that each successive model corrects the predictions of its predecessor, ultimately resulting in a highly accurate meta-model. There are four major categories of boosting algorithms: adaptive boosting, cat boosting, gradient boosting, and light gradient boosting. This research utilized adaptive boosting (AdaBoost) as one of the boosting algorithms.

8.3.4.1 Random Forest

Random Forest is an ensemble method that combines multiple randomized DTs, and their predictions are aggregated using an averaging technique [33]. A DT is a simple hierarchical algorithm used mainly for classification. It uses a tree-like structure with root nodes, branches, and leaf nodes to predict a target variable from input variables. The tree is built by splitting the dataset at the root node into subsets, a process that repeats until a stopping condition is reached. The RF is highly adaptable and is well-suited for tackling extensive challenges, particularly when dealing with high-dimensional feature spaces and intricate data structures in computational biology. Using random sampling and ensemble techniques in RF contributed to improved prediction and generalization, effectively addressing the issues related to DT, such as variance. The algorithm is constructed based on weak learners, specifically DT, creating an analogy of using trees to form a forest.

8.3.4.2 Extremely Randomized Trees

Highly Randomized Trees or Extra Trees [34] is another ensemble method that operates in the same way as Random Forest, but with a distinction in how the decision trees within the forest are built. The algorithm trains decision trees on the entire dataset without using bootstrapped samples. It randomly chooses values to split a feature and generate child nodes. When splitting tree nodes, the tree-based algorithm incorporates generalization in its decisions for attributes and cut-points. The true power of randomization becomes evident when the right parameter values are chosen. The primary benefit of using this tree-based ensemble is its computational efficiency.

8.3.4.3 AdaBoost

AdaBoost forms a strong learner by combining decision trees, referred to as weak learners, which are designed with just one division, referred to as decision stumps (DS) [35]. In the initial creation of the first decision stump in AdaBoost, all data points are given equal weightage. To address previous mistakes, data points that were misclassified previously carry more weight than those that were classified correctly. This process is executed for a defined number of iterations or until the intended results are attained. The working algorithm of AdaBoost can be summed up in the following steps:

1. Allocate uniform weights to all instances M in the dataset, following Equation 8.6:
$$Z(n_j, g_j) = \frac{1}{M}, \quad j = 1, 2, \ldots, m \tag{8.6}$$
2. For each feature, a decision stump (DS) is generated, and the Gini Index (GI) is computed using Equation 8.7:
$$GI = 1 - (\text{probability of true})^2 - (\text{probability of false})^2 \tag{8.7}$$
3. The tree with the lowest GI is selected as the initial decision stump. The importance of the created DS is determined using Equations 8.8 and 8.9:
$$\begin{aligned}\text{Sample weight for incorrectly classified data points} = \\ \text{old sample weight} \times \exp(\text{amount of say})\end{aligned} \tag{8.8}$$
$$\begin{aligned}\text{Sample weight for correctly classified data points} = \\ \text{old sample weight} \times \exp(-\text{amount of say})\end{aligned} \tag{8.9}$$
4. Compute the overall error and performance to update weights. Repeat the process iteratively until the desired results are achieved.

The training phase of the AdaBoost model involved configuring it on the training dataset, using specific hyperparameters such as the number of decision trees (`n_estimators`), the maximum depth of decision trees (`max_depth`), and the learning rate (`learning_rate`). At first, the model underwent training with the default values of chosen hyperparameters, which were subsequently optimized to enhance the performance of AdaBoost. The best set of hyperparameters was determined through a grid search cross-validation process that thoroughly explores the hyperparameter space (various possible combinations of hyperparameter values) to identify the most effective configuration.

8.4 RESULTS

The critical aspect of developing a predictive model for classifying MHC binders is whether the developed model can distinguish binding peptides from the non-binding ones with reliable accuracy. Thus it becomes of prime importance to evaluate the developed system using relevant metrics. An important tool that provides the most plain and concise way of visualizing and interpreting the model's performance is the confusion matrix. It also helps to calculate well established evaluation metrics that provide valuable insights into the different aspects of model's performance. The confusion matrix is a matrix containing the following elements:

1. **True Positive** (**TP**): TPs are the positive instances identified as positive by the model.
2. **True Negative** (**TN**): TNs are the negative instances identified as negative by the model.
3. **False Positive** (**FPs**): FPs are the negative instances identified as positive by the model.
4. **False Negative** (**FN**): FNs are the positive instances identified as negative by the model.

This section explores the outcomes attained by utilizing various models, shedding light on the relative effectiveness of those models and offering a thorough comprehension of performance among the utilized algorithms.

8.4.1 EVALUATION METRICS

In this study, the following metrics derived from the confusion matrix have been used for evaluating efficiency:

1. **Accuracy**: Accuracy provides a straightforward performance assessment of the model by simply measuring the correct predictions against the total number of predictions. Mathematically, accuracy can be calculated using Equation 8.10.

$$\text{Accuracy} = \frac{TP+TN}{TP+TN+FP+FN} \tag{8.10}$$

2. **Precision**: Precision quantifies the number of correct positive predictions against all positive predictions. In case of our study, minimizing false positives is fundamental as they may divert the attention from potential inducing peptides. Precision can be calculated using the formula given in Equation 8.11.

$$\text{Precision} = \frac{TP}{TP+FP} \tag{8.11}$$

3. **Recall**: Also known as Sensitivity, this metric determines the rate at which the model can identify positive instances among all instances. This is particularly important in this study, as a lower recall rate may lead to the prolonged allocation of resources. Recall can be computed using Equation 8.12.

$$\text{Recall} = \frac{TP}{TP+FN} \tag{8.12}$$

4. **F1 Score**: F1 score combines Precision and Recall and provides a balanced assessment of the model's performance, especially in classification problems that utilize imbalanced data, which is the case in our study. F1 score can be determined using the formula in Equation 8.13.

$$\text{F1 score} = \frac{2 \cdot \text{Precision} \cdot \text{Recall}}{\text{Precision} + \text{Recall}} \tag{8.13}$$

5. **AUROC Score**: In addition to these four commonly used metrics, the model has also been assessed using the AUROC score (Area Under the Receiver Operating Characteristics Curve). It measures the area under the ROC curve and represents the model's ability to distinguish between two classes. This metric is particularly useful in biological datasets, where distinguishing between positive and negative instances is pivotal. Mathematically, the AUROC score can be calculated using Equation 8.14.

$$\text{AUROC} = \int \text{TPR}\, d(\text{FPR}) \tag{8.14}$$

8.4.2 PERFORMANCE ANALYSIS

Multiple models were trained and tested to identify the one that most effectively comprehends the relationship between input features and accurately classifies the peptides in question. As mentioned earlier, models like Logistic Regression and SVM based models exhibited a low degree of accuracy, as indicated in Table 8.2. From the Table 8.2, the superior performance of tree-based models can be observed among which the Ensemble model of DTs i.e the AdaBoost model is outperforming all. This lead to the strategic decision of employing AdaBoost as the final model.

The distribution of various performance metrics across different hyperparameter values of the AdaBoost Classifier has been presented in the form of various plots and tables. From Figure 8.10, an increasing trend in accuracy can be observed with the

TABLE 8.2
Progressive Accuracy Trend across Various Models

Models	Accuracy
Logistic Regression	0.61
SVM	0.62
Kernel SVM	0.66
Decision Tree	0.67
Random Forest	0.69
Extra Tree	0.70
AdaBoost	0.72

number of DTs employed in the training of AdaBoost. Figure 8.10 suggests that the model benefits from adding more weak learners (individual DTs). Consequently, a maximum number of DTs were employed in the training of the final model. A similar trend can be observed in Figure 8.11 where the accuracy of the classifier is increasing with respect to the increasing depth of DTs employed in its training. The upward shift of accuracy observed in Figure 8.11 points to the need to use more complex DTs in the training process of AdaBoost instead of basic Decision Stumps. As the peptide dataset is feature-rich, a more complex model with deeper DTs contributes

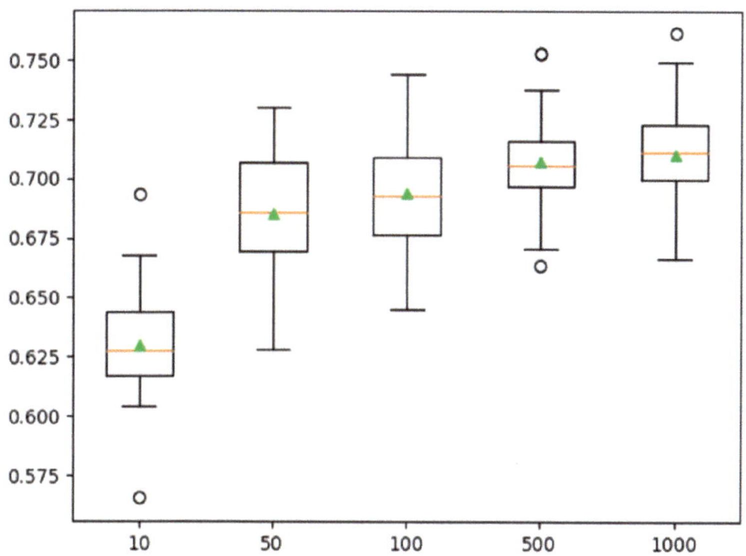

FIGURE 8.10 Accuracy for different no. of DTs.

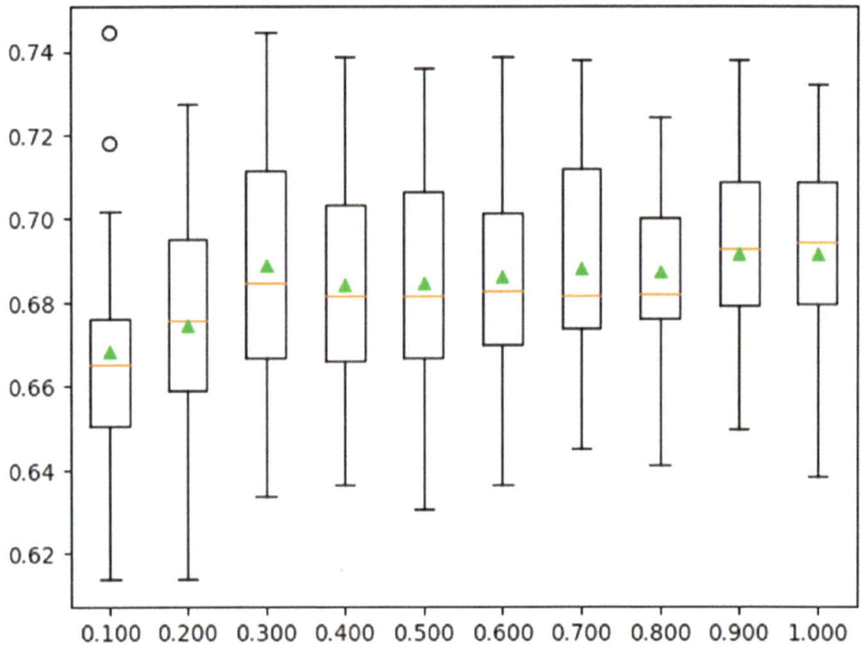

FIGURE 8.11 Accuracy for different values of LR.

positively to the overall performance. Contrary to these two, Figure 8.12 depicts that the proposed model is robust to learning rate (LR) where the accuracy of the classifier remains almost constant despite changing the learning rate from 0.1 to 1.0. Accordingly, the learning rate was retained to maintain the comprehensive integration of weak learners into the AdaBoost.

However, to acquire a more intuitive understanding of the model's performance, it's essential to consider additional performance metrics and their variations with respect to the changing hyperparameters. Table 8.3 lists the values of different accuracy metrics against the number of DTs employed. Table 8.4 provides a record of various accuracy metrics in relation to the depth of DTs chosen. From the Tables 8.3 and 8.4, it can be clearly observed that the performance metrics also demonstrate an increase, as is the case with accuracy. Table 8.5 illustrates the consistency of other metrics, including accuracy, as the learning rate (LR) increases. Finally, the chosen values for various hyperparameters are presented in Table 8.6.

Subsequently, the tuned model achieved an accuracy of 0.81, Precision of 0.84, Recall value of 0.81 and F1 score of 0.82 as enumerated in Table 8.7. Figure 8.13 represents the AUROC of the built model with an AUROC score reaching up to 0.889. These results clearly indicate that it is feasible to categorize MHC-binding peptides with a reasonable level of accuracy using the proposed ML model. An additional

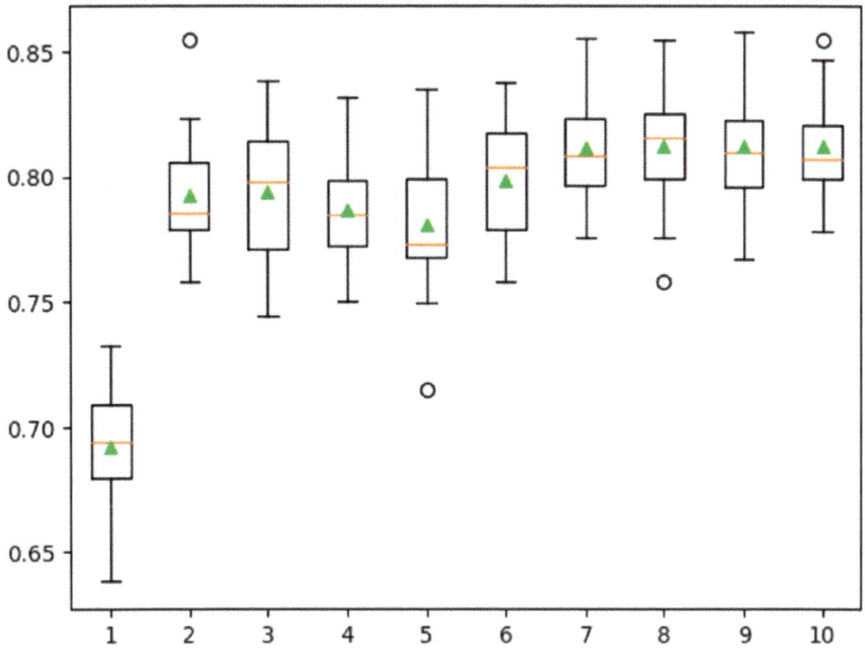

FIGURE 8.12 Accuracy for different depths of DT.

advantage is the ease of implementation, as this model can be executed on a standard personal computer.

8.4.3 COMPARATIVE ANALYSIS

Despite the importance of IL-4 induction in immune responses, scarce literature exists addressing the prediction of IL-4-inducing peptides. The limited work on this

TABLE 8.3
Evaluation Metrics for Different No. of DTs

No. of DTs	Accuracy	Precision	Recall	F1 score
10	0.63	0.65	0.63	0.64
50	0.68	0.66	0.66	0.66
100	0.69	0.68	0.68	0.68
500	0.70	0.70	0.68	0.69
1000	0.71	0.71	0.69	0.70

TABLE 8.4
Evaluation Metrics for Different Depths of DTs

Max Depth	Accuracy	Precision	Recall	F1 score
1	0.69	0.70	0.66	0.67
3	0.79	0.81	0.69	0.74
5	0.78	0.81	0.72	0.76
7	0.81	0.82	0.79	0.80
9	0.81	0.82	0.80	0.81

TABLE 8.5
Evaluation Metrics for Different Values of LR

LR	Accuracy	Precision	Recall	F1 score
0.1	0.67	0.66	0.63	0.64
0.5	0.68	0.67	0.65	0.66
1.0	0.69	0.68	0.66	0.67

TABLE 8.6
Selected Values of Hyperparameters for Final Model

Hyperparameters of AdaBoost	Selected Value
No. of DTs	1000
Learning Rate	1.0
Depth of DT	9

TABLE 8.7
Evaluation Metrics for the Proposed Model

Evaluation Metrics	Corresponding Value
Accuracy	0.81
Precision	0.84
Recall	0.81
F1 score	0.82

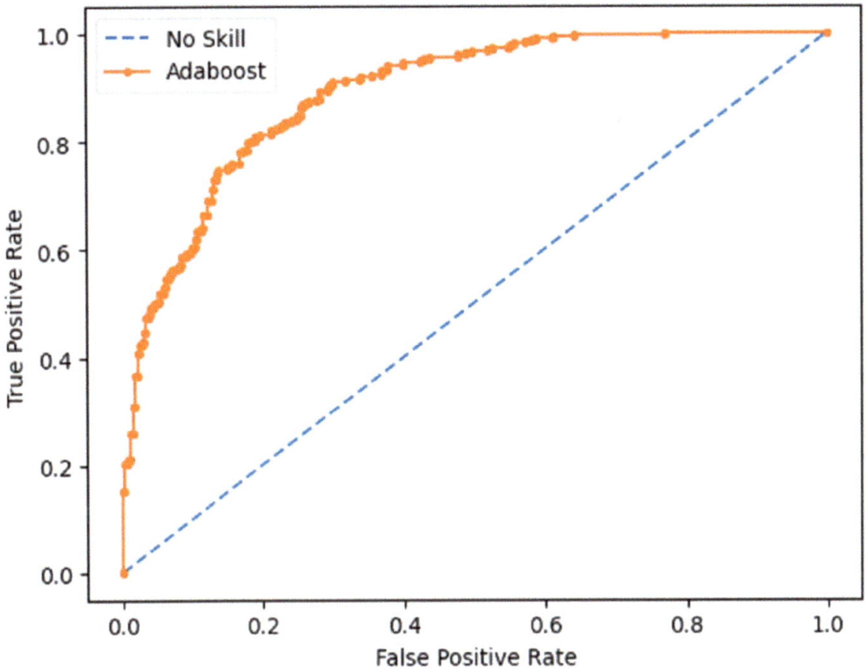

FIGURE 8.13 AUROC curve of the proposed model.

subject necessitates the exploration of ML methodologies to unravel the complexities of peptide-induced immune responses. This research marks progress in bridging the research gap related to the application of ML in distinguishing IL-4-inducing peptides from non-inducing ones. As indicated by Table 8.8, the proposed methodology demonstrates effective comparability with existing approaches in its capacity to categorize peptides. By introducing an ensemble ML approach, this research broadens the range of available methodologies and improves the predictive accuracy. In comparison to an SVM-based approach proposed by [24], our model demonstrates

TABLE 8.8
Comparison between Existing Models and Proposed Approach Using Various Metrics

Author	Prediction Model	Accuracy(%)	AUROC	MCC
Dhanda et al. [8]	SVM	75.76	-	0.51
Proposed Model	AdaBoost	**81.0**	**0.89**	**0.61**

superior accuracy, achieving 81% compared to 75%. Our model demonstrated a commendable AUROC score of 88.9% in a more comprehensive assessment, indicating a superior ability to distinguish between true positive and false positive rates. This outperforms the AUROC score of 80% achieved by IL-13pred developed by [36]. With regard to tree-based Ensemble methods, our Ensemble model achieved an MCC (Mathew's Correlation coefficient) score of 0.61 as compared to a value of 0.59 attained by a Random forest-based model proposed by [26]. This underscores the potential application of ML in developing effective targeted immunization strategies.

8.5 CONCLUSION

The application of advanced ML techniques in the field of computational biology has yielded significant advancements in disease prediction, protein function discovery, genomic sequence analysis, gene discovery, multiple alignment, and phylogeny construction, leading to remarkable outcomes. Classifying IL-4-inducing peptides holds great significance as IL-4 serves multiple essential functions in the body and finds applications in addressing autoimmune diseases, drug development, and immune response modulation. Traditional methods for identifying IL-4-inducing peptides are time-consuming and complex. The recommended model attained a notable level of accuracy, particularly in its ability to predict the positive class, which is noteworthy. This achievement is of great importance, as making incorrect predictions could potentially result in the loss of valuable data that could otherwise be utilized for increasing the efficacy of vaccines. Our model surpassed the existing procedures of identifying the IL-4-inducing peptides with an accuracy rate of 81%. Nonetheless, other pertinent matters could be explored in the future, such as integrating additional features and employing state-of-the-art techniques in ML. The existence of extensive peptide sequence datasets offers the potential to enhance the suggested system, minimizing errors and establishing a dependable detection system. A package of predictive and a feature generation modules has the potential to streamline tasks in this domain, making people more inclined to engage in future endeavours and potentially enhancing overall results.

REFERENCES

1. CA Janeway, P Travers, M Walport, and MJ Shlomchik. Adaptative immunity to infection. *Immunobiology. 5th ed. New York: Churchill Livingstone*, pages 381–423, 2001.
2. Ilona Gutcher, Burkhard Becher, et al. Apc-derived cytokines and t cell polarization in autoimmune inflammation. *The Journal of clinical investigation*, 117(5):1119–1127, 2007.
3. Daria Briukhovetska, Janina Dorr, Stefan Endres, Peter Libby, Charles A Dinarello, and Sebastian Kobold. Interleukins in cancer: from biology to therapy. *Nature Reviews Cancer*, 21(8):481–499, 2021.
4. Kamran Ghoreschi, Peter Thomas, Susanne Breit, Martin Dugas, Reinhard Mailhammer, Willem van Eden, Ruurd van der Zee, Tilo Biedermann, Jorg Prinz, Matthias Mack, et al. Interleukin-4 therapy of psoriasis induces th2 responses and improves human autoimmune disease. *Nature medicine*, 9(1):40–46, 2003.

5. Caterina Musolino, Alessandro Allegra, Vanessa Innao, Andrea Gaetano Allegra, Giovanni Pioggia, Sebastiano Gangemi, et al. Inflammatory and anti-inflammatory equilibrium, proliferative and antiproliferative balance: the role of cytokines in multiple myeloma. *Mediators of inflammation*, 2017, 2017.
6. Erik Lubberts, Liduine van den Bersselaar, Birgitte Oppers-Walgreen, Paul Schwarzenberger, Christina JJ Coenen-de Roo, Jay K Kolls, Leo AB Joosten, and Wim B van den Berg. Il-17 promotes bone erosion in murine collagen-induced arthritis through loss of the receptor activator of nf-κb ligand/osteoprotegerin balance. *The Journal of Immunology*, 170(5):2655–2662, 2003.
7. Dolores Njoku. Suppressive and pro-inflammatory roles for il-4 in the pathogenesis of experimental drug-induced liver injury: A review. *Expert opinion on drug metabolism & toxicology*, 6:519–531, 02 2010.
8. Arvind Kumar Tiwari and Rajeev Srivastava. A survey of computational intelligence techniques in protein function prediction. *International journal of proteomics*, 2014, 2014.
9. Jeff Heaton. Ian goodfellow, yoshua bengio, and aaron courville: Deep learning: The mit press, 2016, 800 pp, isbn: 0262035618. *Genetic programming and evolvable machines*, 19(1-2):305–307, 2018.
10. Monika Khandelwal and Ranjeet Kumar Rout. Prmxai: protein arginine methylation sites prediction based on amino acid spatial distribution using explainable artificial intelligence. *BMC bioinformatics*, 24(1):376, 2023.
11. Ranjeet Kumar Rout, Saiyed Umer, Monika Khandelwal, Smitarani Pati, Saurav Mallik, Bunil Kumar Balabantaray, and Hong Qin. Identification of discriminant features from stationary pattern of nucleotide bases and their application to essential gene classification. *Frontiers in Genetics*, 14:1154120, 2023.
12. Monika Khandelwal, Ranjeet Kumar Rout, and Saiyed Umer. Prediction of protein-protein interaction using support vector machine based on spatial distribution of amino acids. In *International Conference on Advances and Applications of Artificial Intelligence and Machine Learning*, pages 23–32. Springer, 2022.
13. Harpreet Singh and GPS Raghava. Propred1: prediction of promiscuous mhc class-i binding sites. *Bioinformatics*, 19(8):1009–1014, 2003.
14. Sneh Lata, Manoj Bhasin, and Gajendra PS Raghava. Application of machine learning techniques in predicting mhc binders. *Immunoinformatics: Predicting Immunogenicity In Silico*, pages 201–215, 2007.
15. Geng Liu, Dongli Li, Zhang Li, Si Qiu, Wenhui Li, Cheng-chi Chao, Naibo Yang, Handong Li, Zhen Cheng, Xin Song, et al. Pssmhcpan: a novel pssm-based software for predicting class i peptide-hla binding affinity. *Giga Science*, 6(5):gix017, 2017.
16. Morten Nielsen, Claus Lundegaard, Thomas Blicher, Kasper Lamberth, Mikkel Harndahl, Sune Justesen, Gustav Rder, Bjoern Peters, Alessandro Sette, Ole Lund, et al. Netmhcpan, a method for quantitative predictions of peptide binding to any hla-a and-b locus protein of known sequence. *PloS one*, 2(8):e796, 2007.
17. Ranjeet Kumar Rout, Sk Sarif Hassan, Sabha Sheikh, Saiyed Umer, Kshira Sagar Sahoo, and Amir H Gandomi. Feature-extraction and analysis based on spatial distribution of amino acids for sars-cov-2 protein sequences. *Computers in biology and medicine*, 141:105024, 2022.
18. Morten Nielsen and Massimo Andreatta. Netmhcpan-3.0; improved prediction of binding to mhc class i molecules integrating information from multiple receptor and peptide length datasets. *Genome medicine*, 8(1):1–9, 2016.

19. Harpreet Singh and GPS Raghava. Propred: prediction of hla-dr binding sites. *Bioinformatics*, 17(12):1236–1237, 2001.
20. Manoj Bhasin and GPS Raghava. Svm based method for predicting hla-drb1* 0401 binding peptides in an antigen sequence. *Bioinformatics*, 20(3):421–423, 2004.
21. Morten Nielsen and Ole Lund. Nn-align. an artificial neural network-based alignment algorithm for mhc class ii peptide binding prediction. *BMC bioinformatics*, 10(1):1–10, 2009.
22. Anjali Dhall, Sumeet Patiyal, and Gajendra PS Raghava. Hlancpred: a method for predicting promiscuous non-classical hla binding sites. *Briefings in Bioinformatics*, 23(5):bbac192, 2022.
23. Sneh Lata and GPS Raghava. Cytopred: a server for prediction and classification of cytokines. *Protein Engineering, Design & Selection*, 21(4):279–282, 2008.
24. Sandeep Kumar Dhanda, Sudheer Gupta, Pooja Vir, and GPS Raghava. Prediction of il4 inducing peptides. *Clinical and Developmental Immunology*, 2013, 2013.
25. Anjali Dhall, Sumeet Patiyal, Neelam Sharma, Salman Sadullah Usmani, and Gajendra PS Raghava. Computer-aided prediction and design of il-6 inducing peptides: Il-6 plays a crucial role in covid-19. *Briefings in bioinformatics*, 22(2):936–945, 2021.
26. Gandharva Nagpal, Salman Sadullah Usmani, Sandeep Kumar Dhanda, Harpreet Kaur, Sandeep Singh, Meenu Sharma, and Gajendra PS Raghava. Computer-aided designing of immunosuppressive peptides based on il-10 inducing potential. *Scientific reports*, 7(1):42851, 2017.
27. Sudheer Gupta, Parul Mittal, Midhun K Madhu, and Vineet K Sharma. Il17escan: a tool for the identification of peptides inducing il-17 response. *Frontiers in immunology*, 8:1430, 2017.
28. Shipra Jain, Anjali Dhall, Sumeet Patiyal, and Gajendra PS Raghava. Il13pred: A method for predicting immunoregulatory cytokine il-13 inducing peptides for managing covid-19 severity. *bioRxiv*, pages 2021–2040, 2021.
29. Randi Vita, Swapnil Mahajan, James A Overton, Sandeep Kumar Dhanda, Sheridan Martini, Jason R Cantrell, Daniel K Wheeler, Alessandro Sette, and Bjoern Peters. The immune epitope database (iedb): 2018 update. *Nucleic acids research*, 47(D1):D339–D343, 2019.
30. Amos Bairoch and Rolf Apweiler. The swiss-prot protein sequence database and its supplement trembl in 2000. *Nucleic acids research*, 28(1):45–48, 2000.
31. Nitesh V Chawla, Kevin W Bowyer, Lawrence O Hall, and W Philip Kegelmeyer. Smote: synthetic minority over-sampling technique. *Journal of artificial intelligence research*, 16:321–357, 2002.
32. Thomas G Dietterich. Ensemble methods in machine learning. In *International workshop on multiple classifier systems*, pages 1–15. Springer, 2000.
33. Adele Cutler, D Richard Cutler, and John R Stevens. Random forests. *Ensemble machine learning: Methods and applications*, pages 157–175, 2012.
34. Pierre Geurts, Damien Ernst, and Louis Wehenkel. Extremely randomized trees. *Machine learning*, 63:3–42, 2006.
35. Yoav Freund, Robert E Schapire, et al. Experiments with a new boosting algorithm. In *icml*, volume 96, pages 148–156. Citeseer, 1996.
36. Shipra Jain, Anjali Dhall, Sumeet Patiyal, and Gajendra PS Raghava. Il13pred: A method for predicting immunoregulatory cytokine il-13 inducing peptides. *Computers in Biology and Medicine*, 143:105297, 2022.

9 Healthcare Portal-Django Framework for Healthcare Management System

Kamalpreet Kaur, Monika Khandelwal, and Renu Dhir

9.1 INTRODUCTION

Within the context of the modern healthcare system, the efficient administration of hospital operations is of the utmost importance to guarantee that patients receive high-quality medical attention [1]. It has become clear that one of the most critical steps in this direction is the shift away from traditional manual procedures and toward digital platforms. The quest to streamline hospital management has reached a critical milestone with the completion of the project dubbed "Hospital Management System in Django." Within the scope of this research paper, an in-depth investigation of the system is presented, with particular emphasis placed on its architecture, functionality, and the numerous advantages it provides [2, 3].

The fundamental objective of the Hospital Management System is to produce a user-friendly, quick, and cost-effective software package. Its primary aim is to automate operations and limit the amount of manual intervention that is required [4, 5]. The Hospital Management System intends to revolutionize the Front Office Management of hospitals. The system has been methodically built to manage extensive duties, including medical history, appointment scheduling, patient and doctor information, and administrative responsibilities. This system emphasizes digitalization and provides four unique dashboards for patients, doctors, receptionists, and HR staff. Each dashboard is supplied with individualized access and functionality concerning the user's specific needs [6].

This article aims to investigate the numerous functions included in the Hospital Management System (Figure 9.1) [7]. These functions include patient registration, appointment scheduling, management of invoices and payments, and the creation of prescriptions. In addition, it will ensure that the system can maintain medical records securely and efficiently. Additionally, the functions and tasks of

DOI: 10.1201/9781032660714-9

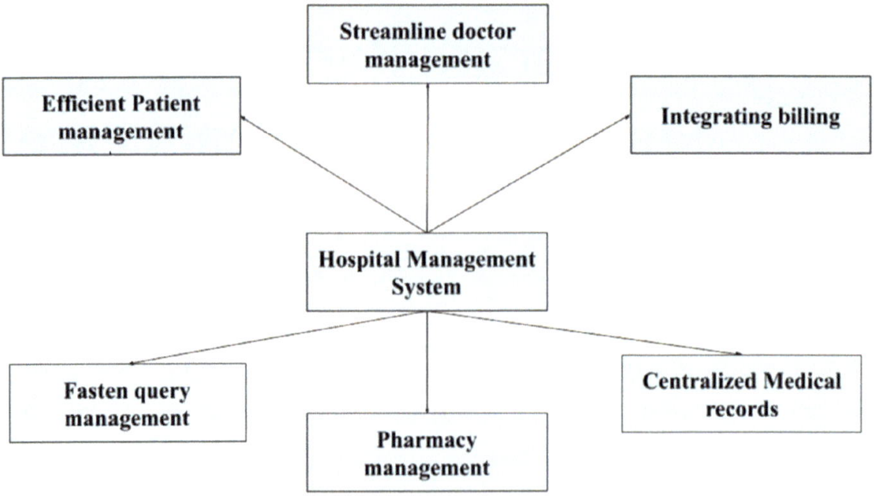

FIGURE 9.1 Features of the hospital management system.

the four unique dashboards, which are the patient dashboard, the doctor dashboard, the receptionist dashboard, and the HR staff dashboard, will be explained in detail [8].

The Hospital Management System is geared to manage the comprehensive digitization of hospital operations. It involves registering patients, storing their details in the system, and the electronic tracking of pharmacy and lab activities [9]. The software package generates a unique ID for each patient and automatically maintains the details of every patient and employee. It features a groundwork facility to understand the status of each region. Users can easily search for the availability of a doctor and get patient details using the unique ID [10]. A username and password are required to access the system and are available to the administrator or secretary. Only authorized individuals can add data to the system, maintaining data security. The data retrieval method is straightforward and efficient. The interface is highly user-friendly, and the data are safeguarded for personal use, ensuring rapid and secure information processing [1]. The Hospital Management System is a robust, versatile, and easy-to-use platform created and developed to bring real and considerable benefits to hospitals. By offering a digital foundation, this system improves the daily operations of hospitals, assuring efficient patient care and optimized resource usage [3].

9.2 SCOPE OF THE SURVEY

This chapter aims to undertake a detailed examination of the "Hospital Management System in Django" project. The survey will cover several facets of the system, including its architecture, functionality, and the benefits it gives. Key areas to be explored include:

A. System Architecture

1. Understand the general structure and components of the Hospital Management System.
2. Analyze the interaction between different system components.
3. Examine the integration of Django, Python, and SQLite within the system.

B. Functionality

1. Explore core features such as patient registration, appointment scheduling, and medical history management.
2. Assess the efficiency and effectiveness of the system in managing patient and doctor information, appointment scheduling, prescription creation, invoice generation, and payment administration.
3. Review the automation components integrated into the pharmacy and laboratory processes.

C. User Access and Roles

1. Study the four unique dashboards designed for patients, doctors, receptionists, and HR professionals.
2. Analyze the responsibilities and access levels assigned to each user type within the system.
3. Evaluate the user interface and the ease of navigation for each dashboard.

D. Data Security and Protection

1. Investigate the security measures implemented to ensure the integrity and confidentiality of patient and hospital data [11].
2. Assess the system's ability to protect sensitive medical records and personal information.

E. Advantages and Benefits

1. Identify the advantages and benefits of implementing the Hospital Management System.
2. Analyze the system's impact on hospital operations, resource optimization, and patient care.

3. Examine how the system streamlines hospital workflows and enhances patient care efficiency.

F. Future Additions

1. Discuss potential future enhancements, such as incorporating blockchain and AI, to improve functionality, data security, and automation [1, 12].
2. Propose strategies for the continuous improvement and development of the Hospital Management System.

The survey provides an in-depth insight into the Hospital Management System in Django, helping stakeholders fully appreciate its capabilities, benefits, and future potential.

9.3 METHODOLOGY

The technique followed in the creation and evaluation of the Hospital Management System (HMS) in Django is explained as follows:

1. **System Development Framework:** The Hospital Management System (HMS) in Django was designed utilizing the Django framework. Django features a pragmatic and clear design, making it a good alternative for rapid development. Additionally, it enforces security rules well [9].
2. **Programming Languages:** The system was designed using Python, a high-level and interpreted language well-suited for quick development. SQLite, a C-language library, was utilized as the database, being lightweight and requiring minimal setup.
3. **Development of User Interfaces:** Four separate user interfaces were developed: Patient, Doctor, Receptionist, and HR interfaces. This technique enabled role-based access, ensuring specialized functionality for each group of users [8].
4. **Data Collection and Processing:** Data regarding patients, doctors, appointments, prescriptions, invoices, and payments were collected and handled efficiently utilizing Django models. This methodology allowed creating, querying, and modifying data.
5. **Security Implementation:** The security of the system was secured by using Django's authentication system, safeguarding the data and managing user rights. Furthermore, encrypted communication routes were built to avoid data breaches and unwanted access [2].
6. **Testing and Quality Assurance:** Rigorous testing procedures, including unit testing and system integration testing, were applied to assure the dependability, stability, and efficacy of the system.
7. **Comparative Analysis with Competitors:** Existing healthcare management systems were critically examined based on functionality, performance, scalability, and ease of use. This analysis helped highlight the advantages and drawbacks of the Hospital Management System in Django.

8. **Future Enhancements Integration:** The potential future advancements, including Blockchain and Artificial Intelligence integration, were recognized, and the tactics to include them were reviewed to ensure the system remains technologically sophisticated.

9.4 LITERATURE REVIEW

In 2005, Hillestad et al. [13] explored the potential health and financial benefits of electronic medical record (EMR) systems, estimating annual savings of over $81 billion through improved efficiency and safety, yet highlighting barriers like high costs and lack of standardization requiring system-wide changes for realization. Lenz et al. [8] examined IT assistance for healthcare processes, focusing on medical knowledge evolution and distributed healthcare networks. Promotes innovative process management technology to increase healthcare quality and IT flexibility. Hardeep Singh [14] examined strategies for involving patients in reducing diagnostic errors, emphasizing their role in their own care, providing feedback to the healthcare system, and contributing to research and policy development on diagnosis-related issues. Additionally, Engel et al. [4] examined challenges and solutions in demand and supply for diagnostics. Findings highlight operational barriers like logistics and cost, proposing solutions, such as health system strengthening and supply chain innovations. Pourasghar et al. [6] identified communication barriers in patient transfer, including inadequate medical history and incomplete documents.

Consequences include managerial, clinical, economic, and social impacts, underscoring the need for improved protocols, monitoring, and inter-sectoral collaboration [6]. Verna et al. [9] explored strategies for reducing diagnostic errors and increasing efficiency through diagnostic management teams. It advocates for improved communication between healthcare providers and diagnosticians, emphasizing the global need for accurate and rapid diagnoses as diagnostic information becomes more complex and costly. Ayesha et al. [12] explored the potential of blockchain technology in transforming Electronic Health Record (EHR) systems. It proposes a framework for secure record storage and granular access rules, addressing the integrity and security of data. Future work includes implementing a payment module within the framework to enhance healthcare service delivery. Kubendiran et al. [2] proposed a framework that ensures data integrity and provenance, crucial for maintaining the confidentiality and reliability of healthcare data by leveraging blockchain technology. The proposed partition-based device discovery scheme further enhances the efficiency and reliability of data discovery in shared controlled networks, offering significant improvements in performance compared to existing schemes.

Liu et al. [1] proposed a blockchain-based Electronic Medical Records Management System to efficiently store and share medical records while ensuring security and privacy. They discussed using a permissioned blockchain platform and outlined the future of testing the system with real patient data and conducting performance testing. Melander et al. [10] proposed an AI diagnostic system to improve healthcare delivery in underserved regions, addressing hindrances and emphasizing the critical priority of accessible diagnostics, advocating for collaboration to develop cost-effective, reliable solutions for global health improvement. Senbekov et al. [7] discussed recent advancements in blockchain platforms, smart devices, telemedicine, big data, artificial intelligence, and other digital health technologies [15] highlighting their impact on healthcare accessibility, quality, and flexibility. Sriram et al. [3] discussed the impact of the Internet of Things (IoT) [16], Omics revolution, and artificial intelligence on the future of health care delivery in the United States. It introduces The P9 health care model, presents a framework for smart health care, and outlines ongoing research within NIST, the National Institute of Standards and Technology. Ali et al. [11] presented a comprehensive framework for securing digital healthcare systems using blockchain technology. The proposed framework provides comprehensive protection for Electronic Health Records (EHR) by incorporating novel security algorithms and leveraging permissioned blockchain networks. Filip et al. [17] reviewed the global challenges faced by public healthcare systems during the COVID-19 epidemic; special attention was paid to the need for reorganization and the adoption of telemedicine and mobile health apps to streamline medical care and prevent future outbreaks. Table 9.1 presents a literature review framework for conducting a comparative analysis of digital advancements in existing healthcare systems.

9.5 RESEARCH GAPS AND CHALLENGES

In the rapidly developing field of healthcare informatics, personalized medicine recommendations are generated through the integration of machine learning (ML) and artificial intelligence (AI) in medicine recommendation systems. There are still a number of research gaps in spite of tremendous advancements.

Even with significant technological developments, there are still a number of enduring problems in the healthcare system that make it difficult to diagnose and treat patients effectively [20]. Information flow between patients, laboratories, and healthcare providers is hampered by fragmented medical records, communication barriers, and a lack of interoperability. Concerns over data security and privacy have increased as healthcare data becomes more digitally connected, underscoring the necessity of secure database solutions [21].

To guarantee the seamless operation and scalability of the proposed architecture, technological problems and compatibility issues arising from the integration of numerous technologies, such as blockchain, Node.js, SQL, and React.js, must be resolved. Additionally, there are still a lot of diagnostic blunders and inefficiencies in the healthcare system, which emphasizes how important it is to give clinicians access to thorough patient health information and tools for decision support in order to increase the accuracy and efficiency of diagnoses [22].

Unfair healthcare disparities are made worse by inadequate access to healthcare services, which makes patient diagnosis and treatment more difficult, particularly in developing nations [23]. Collaboration, feedback, and innovation must continue in order to overcome these challenges and clear the way for the development of a more patient-centered, efficient, and accessible healthcare system. Hospital management systems' successful deployment and continued improvement depend on addressing these research gaps and obstacles [24].

TABLE 9.1
Comparative Analysis of Digital Advancement in Health Care

Year	Reference	Framework	Description	Remarks
2005	[13]	Comparison of health care IT utilization with other industries; estimates EMR adoption savings, costs, and advantages	Looks at the possible financial and health benefits of using electronic medical record (EMR) systems widely	EMR implementation might save approximately $81 billion annually, potentially tripling with HIT-enabled chronic illness prevention and management, but high prices and lack of interoperability are obstacles
2007	[8]	A socio-technical viewpoint drawn from experience and literature	Considers organizational and medical treatment processes while using IT for healthcare. Discusses how IT may improve decision- making, data quality, monitoring, and compliance with medical norms.	Emphasizes being aware of the differences in healthcare processes. explains the function of IT in data quality, rule compliance, and decision support.
2015	[4]	Qualitative content study of an online discussion with 99 academics, policymakers, manufacturers, and healthcare practitioners to identify LMIC diagnostic access constraints and solutions.	Harvard University's Global Health Delivery Project (GHD) held an online conversation on diagnostics demand and supply in low- and medium-income countries (LMICs). This qualitative content analysis discusses the difficulties and solutions.	Highlights operational and health system hurdles rather than technological factors that hinder diagnostic accessibility and affordability in LMICs. Solutions include health system strengthening, innovation, enhanced recommendations, supply chain improvements, and quality assurance.

(*Continued*)

TABLE 9.1 (*Continued*)
Comparative Analysis of Digital Advancement in Health Care

Year	Reference	Framework	Description	Remarks
2016	[6]	Qualitative study using Focus Group Discussions and interviews.	Examines obstacles to information sharing and communication during patient transfers, with a special emphasis on CTGI, referring, and receiving hospitals.	Highlights the necessity of technological utilization, teamwork, protocol reviews, and monitoring to enhance patient transfer communication.
2019	[2]	Design and implementation of the proposed security framework leveraging blockchain technology.	Suggests a improved security framework for e- health systems based on blockchain technology.	Outlines an innovative use of blockchain technology to guarantee provenance and integrity of data in e-health systems.
2020	[1]	The BDL-IBS system's design and implementation make use of distributed ledger technology and blockchain.	Proposes Improved Biomedical Security using Distributed Ledger-based Blockchain Technology (BDL-IBS) to improve healthcare data privacy and security.	Shows how blockchain and distributed ledger technology can improve healthcare trust, privacy, and security, notably in EHR management.
2020	[10]	Examines existing initiatives in providing healthcare services to underprivileged communities, investigates obstacles, and suggests a diagnostic system.	Suggests a framework and process for creating an AI diagnostic system to enhance the provision of healthcare in underprivileged areas.	Focuses on accurate diagnostics in global health, presents a strategy to solve healthcare difficulties in underserved locations, highlights possible impact on healthcare outcomes, and encourages global health agency collaboration.

2020	[7]	Current developments and uses of digital technology in medical fields.	Reviewing recent progress on digital health technologies	The recent pandemics have led to a significant demand for digital health technologies, contributing to improved healthcare accessibility and flexibility.
2021	[11]	Creation and deployment of an innovative blockchain-based system that combines access control and IoT for safe patient health record administration.	Suggests enhancing the security, privacy, and dependability of patient health records by implementing an architecture for digital healthcare systems based on blockchain.	Introduces a blockchain-based algorithm and architecture for digital healthcare security and privacy. The suggested system outperforms blockchain-based alternatives in security, reliability, and authentication.
2022	[18]	Analytical research assessing the legal and regulatory framework controlling AI in healthcare, as well as discussions about cybersecurity concerns and AI's role in healthcare fraud prevention.	Investigates the issues of data security and privacy in India's AI-powered healthcare systems, with a focus on legal issues, government laws, and the consequences of cyber-attacks.	To maximize the benefits of AI in healthcare while protecting patient privacy and data security, India's AI-driven healthcare systems need strong legal and ethical frameworks, updated regulations, and infrastructure investments.
2024	[19]	Django and Python information for building websites.	Django adheres to the Model-View-Controller (MVC) architectural pattern. Python's rapid growth can be attributed to its superior flexibility, security, high performance, ease of use, and simple syntax when compared to PHP language.	Python and Django together can meet the technical requirements for a good website.

9.6 PROPOSED MODEL

Our approach seeks to transform hospital front office administration by developing a quick, easy-to-use, and reasonably priced software solution. This system makes it easier for patients to register, store their information, and handle billing for laboratory and pharmacy services. Every patient's ID is created specifically for them, and all patient and staff information is automatically stored. The system has a dashboard feature that allows you to monitor each department's status. With the ID, users can quickly look up a doctor's availability and a patient's information. An administrator or a secretary can access the Hospital Management System with a username and password. The database can only contain data they add, and retrieving it is simple. The user-friendly interface and well-protected data for private usage ensure quick data processing. Figure 9.2 illustrates the workflow diagram of essential processes in the hospital management system.

1. Our suggested approach is strong, adaptable, and simple to use; it was created with hospitals' actual, practical benefits in mind. Four primary user roles are the model's emphasis: patients, doctors, receptionists, and HR. Every user role has unique capabilities and access.
2. Patients can sign up, log in, and access appointments, payments, and invoicing. They can view medical history, including doctor-prescribed medications, and create dynamic invoices in PDF format with comprehensive information. Patients' profiles can be edited and managed as well.
3. Physicians can sign up, log in, and access patient appointments. In addition to viewing their most recent prescriptions, they can write new ones. A doctor's profile can be edited and managed as well.
4. Receptionists can view completed, upcoming, and total appointments by logging in. They can schedule appointments, complete with time and date, between physicians and patients. Receptionists can update appointments (completed or pending) and paid and unpaid amounts. Patients can also be created with all the necessary information, balances owed, and amounts paid.
5. HR can log in and see the number of patients, doctors, and doctors on duty. Doctor profiles, including department, attendance, pay, and status (active/inactive), can be customized by them. HR may remove doctors from the system. Accounting allows them to view consultation data, create dynamic invoices, and click a button to send emails with the amount of information to patients. The system can be well protected by integrating blockchain and other security algorithms.

9.6.1 RESULT ANALYSIS

Compared to current healthcare administration systems, our suggested model has several benefits. Users with different degrees of technical competence can easily access and use it because of its user-friendly interface. It provides a one-stop shop for all hospital administration requirements, covering various functions such as

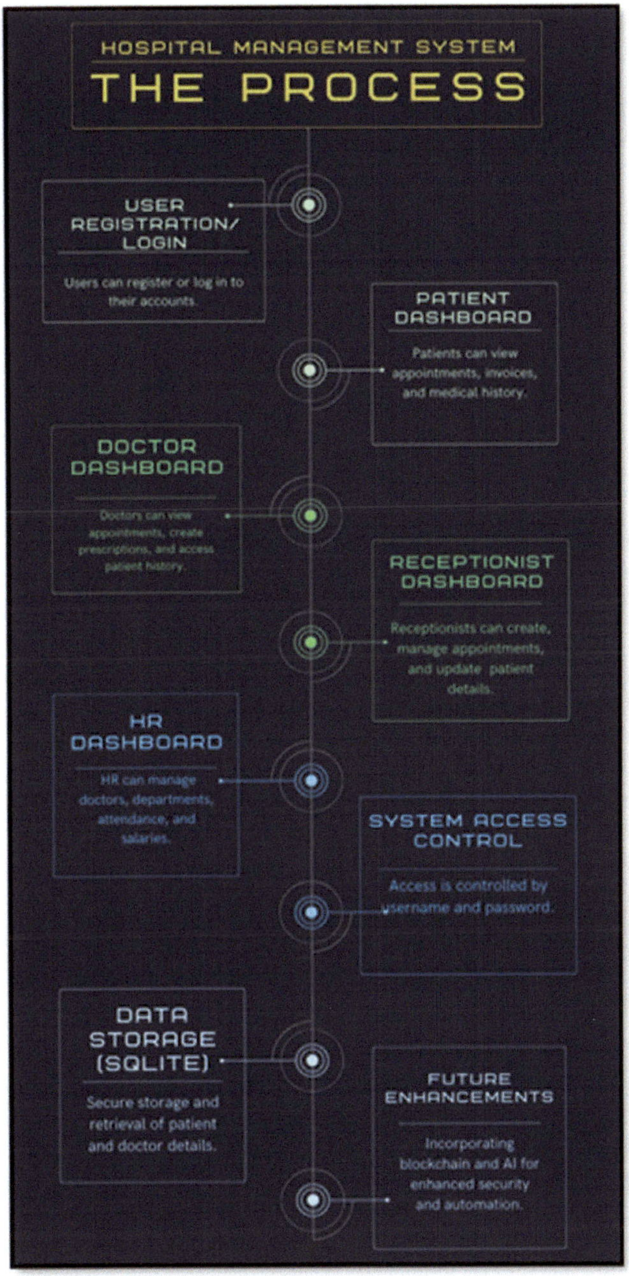

FIGURE 9.2 Flow of the process in the hospital management system.

TABLE 9.2
Comparing the Proposed Model with the Existing Model of Hospital Management System

Parameters	User-Friendly Interface	Efficiency	Scable Architecture
[3]	Low	Low	Yes
[11]	Moderate	Moderate	No
[13]	Less	Moderate	No
[18]	Moderate	Moderate	Yes
[19]	Moderate	Moderate	Yes
Proposed model	High	High	Highly

prescription management, appointment scheduling, patient registration, and invoicing. By incorporating blockchain technology, worries about illegal access and data breaches are allayed, and patient data is kept private and secure. Because of its scalable architecture, the system can meet the requirements of hospitals of different sizes. Additionally, it is adaptable enough to change with the demands and specifications of healthcare facilities. The technology improves the overall efficiency of hospital management by automating a number of activities, including invoicing and appointment scheduling, freeing up medical staff members to concentrate more on patient care. The system offers hospitals and healthcare organizations a high return on investment and is a cost-effective alternative to conventional systems. Streamlining several administrative procedures and encouraging more effective communication between patients and medical staff improves patient care. The various parameters according to the research are well stated in Table 9.2.

9.7 CONCLUSION

The "Hospital Management System" provides a thorough approach to streamlining hospital front office administration. This system simplifies the operations of patient registration, appointment scheduling, prescription management, and invoicing by integrating cutting-edge technologies with an intuitive interface. Its ease of use, extensive functionality, data security, scalability, affordability, and potential to improve patient care set it apart from conventional models. Our suggested model seeks to transform hospital management systems by tackling enduring problems in the healthcare setting and guaranteeing effectiveness, accessibility, and better healthcare delivery. It will need continued innovation, feedback, and collaboration to improve this system in the future and meet the changing demands of the healthcare industry.

REFERENCES

1. Haibing Liu, Rubén González Crespo, and Oscar Sanjuán Martínez. Enhancing privacy and data security across healthcare applications using blockchain and distributed ledger concepts. In *Healthcare*, volume 8, page 243. MDPI, 2020.
2. Mohan Kubendiran, Satyapal Singh, and Arun Kumar Sangaiah. Enhanced security framework for e-health systems using blockchain. *Journal of Information Processing Systems*, 15(2):239–250, 2019.
3. Ram D Sriram and Eswaran Subrahmanian. Transforming health care through digital revolutions. *Journal of the Indian Institute of Science*, 100(4):753–772, 2020.
4. Nora Engel, Keri Wachter, Madhukar Pai, Jim Gallarda, Catharina Boehme, Isabelle Celentano, and Rebecca Weintraub. Addressing the challenges of diagnostics demand and supply: insights from an online global health discussion platform. *BMJ Global Health*, 1(4):e000132, 2016.
5. Monika Khandelwal, Sabha Sheikh, Ranjeet Kumar Rout, Saiyed Umer, Saurav Mallik, and Zhongming Zhao. Unsupervised learning for feature representation using spatial distribution of amino acids in aldehyde dehydrogenase (aldh2) protein sequences. *Mathematics*, 10(13):2228, 2022.
6. Faramarz Pourasghar, Nesa Kavakebi, Jafar Sadegh Tabrizi, and Ahad Banagozar Mohammadi. Barriers to communication and information exchange in patient transfer and its consequences. *Global Journal of Health Science*, 8(12):178, 2016.
7. Maksut Senbekov, Timur Saliev, Zhanar Bukeyeva, Aigul Almabayeva, Marina Zhanaliyeva, Nazym Aitenova, Yerzhan Toishibekov, Ildar Fakhradiyev, et al. The recent progress and applications of digital technologies in healthcare: a review. *International journal of telemedicine and applications*, 2020, 2020.
8. Richard Lenz and Manfred Reichert. It support for healthcare processes–premises, challenges, perspectives. *Data & Knowledge Engineering*, 61(1):39–58, 2007.
9. Roberto Verna, Adriana Berumen Velazquez, and Michael Laposata. Reducing diagnostic errors worldwide through diagnostic management teams. *Annals of Laboratory Medicine*, 39(2):121, 2019.
10. David Melander, Sudesh Sivarasu, Ibrahim Yekinni, Cheng Yunzhang, and Arthur Erdman. Diagnostics as the key to advances in global health: Proposed methods for making reliable diagnostics widely available. *Journal of medical devices*, 14(1):014702, 2020.
11. Aitizaz Ali, Hasliza A Rahim, Muhammad Fermi Pasha, Rafael Dowsley, Mehedi Masud, Jehad Ali, and Mohammed Baz. Security, privacy, and reliability in digital healthcare systems using blockchain. *Electronics*, 10(16):2034, 2021.
12. Ayesha Shahnaz, Usman Qamar, and Ayesha Khalid. Using blockchain for electronic health records. *IEEE access*, 7:147782–147795, 2019.
13. Richard Hillestad, James Bigelow, Anthony Bower, Federico Girosi, Robin Meili, Richard Scoville, and Roger Taylor. Can electronic medical record systems transform health care? potential health benefits, savings, and costs. *Health affairs*, 24(5):1103–1117, 2005.
14. Hardeep Singh. Diagnostic errors: moving beyond 'no respect'and getting ready for prime time, 2013.
15. Anay Ghosh, Bibhas Chandra Dhara, Chiara Pero, and Saiyed Umer. A multimodal sentiment analysis system for recognizing person aggressiveness in pain based on textual and visual information. *Journal of Ambient Intelligence and Humanized Computing*, 14(4):4489–4501, 2023.

16. Alamgir Sardar, Saiyed Umer, Ranjeet Kumar Rout, Kshira Sagar Sahoo, and Amir H Gandomi. Enhanced biometric template protection schemes for securing face recognition in iot environment. *IEEE Internet of Things Journal*, 2024.
17. Roxana Filip, Roxana Gheorghita Puscaselu, Liliana Anchidin-Norocel, Mihai Dimian, and Wesley K Savage. Global challenges to public health care systems during the covid-19 pandemic: a review of pandemic measures and problems. *Journal of personalized medicine*, 12(8):1295, 2022.
18. S Chintala. Data privacy and security challenges in ai-driven healthcare systems in india. *Journal of Data Acquisition and Processing*, 37(5):2769–2778, 2022.
19. Manoj Sharma, Mohammad Shuaib Khan, and Jitendra Singh. Python & django the fastest growing web development technology. In *2024 IEEE 1st Karachi Section Humanitarian Technology Conference (KHI-HTC)*, pages 1–9. IEEE, 2024.
20. Mohsen Vatandoost and Sanaz Litkouhi. The future of healthcare facilities: how technology and medical advances may shape hospitals of the future. *Hospital Practices and Research*, 4(1):1–11, 2019.
21. Barbara L Filkins, Ju Young Kim, Bruce Roberts, Winston Armstrong, Mark A Miller, Michael L Hultner, Anthony P Castillo, Jean-Christophe Ducom, Eric J Topol, and Steven R Steinhubl. Privacy and security in the era of digital health: what should translational researchers know and do about it? *American journal of translational research*, 8(3):1560, 2016.
22. Gilad J Kuperman. Health-information exchange: why are we doing it, and what are we doing? *Journal of the American Medical Informatics Association*, 18(5):678–682, 2011.
23. Wayne J Riley. Health disparities: gaps in access, quality and affordability of medical care. *Transactions of the American Clinical and Climatological Association*, 123:167, 2012.
24. Ann C Greiner, Elisa Knebel, et al. Challenges facing the health system and implications for educational reform. In *Health Professions Education: A Bridge to Quality*. National Academies Press (US), 2003.

10 Unveiling Genetic Codes: Harnessing Machine Learning and Deep Learning for Deoxyribonucleic Acid Sequence Analysis

Shaik Akbar, A. Pathanjali Sastri, and S.G.S. Priyanka

10.1 INTRODUCTION

The primary focus of cellular genetics and genomic analysis is to apply machine learning (ML) and deep learning (DL) as essential tools to understand the complexities of deoxyribonucleic acid (DNA) ordering and conduct various research activities in this field. DNA sequencing, combined with computational insights, is an expensive process, making it necessary to use ML and DL models to estimate DNA ordering outcomes based on existing datasets. Several studies have demonstrated the effectiveness of ML and DL in genomic research, particularly in gene expression analysis, sequence classification, and variant effect prediction [1]. The integration of ML and DL techniques has significantly enhanced the accuracy and efficiency of genomic data analysis [2, 3]. Since the human genome contains nearly three billion nucleotides, manual sequencing is impractical from both economic and clinical perspectives. ML and DL techniques play a significant role in conducting comprehensive data analysis and extracting relevant information essential for future applications. Since DNA sequencing is a vast field, incorporating advanced technologies from Industry 4.0 is necessary to improve reliability in future predictions. The extensive scope of the medical field offers numerous opportunities for ML-based predictions, particularly in forecasting potential medical conditions. Predicting an individual's future health,

DOI: 10.1201/9781032660714-10

especially in cases where medical issues may not be apparent in initial diagnoses, requires the application of predictive models such as decision tree (DT), random forest (RF), and various DL algorithms [4].

The inherent risks associated with medical sequencing predictions and patient prognosis present challenges that traditional diagnostic methods alone cannot effectively address. Technological advancements are crucial in overcoming these limitations. Several studies have demonstrated the effectiveness of ML and DL in genomic research, particularly in variant calling and genome annotation tasks [3,5]. The identification of specific DNA sequences relies on the categorization used in training prototypes. DL models and specialized ML techniques can analyze DNA data in-depth, examining its components and features to gain a clearer understanding of a patient's health condition. The focus remains on identifying datasets and refining them to enhance their suitability for ML and DL methodologies.

10.2 LITRATURE REVIEW

The paper introduces a model that captures the essential local features of DNA capillary electrophoretic time series, followed by the cubic centimetre sequencing algorithm based on this model. The evaluation of the algorithm's properties involves both simulated and validated data. Current DNA sequencing methods are often heuristic and modest in utilizing statistical information. Here, a comprehensive model of DNA estimation is presented, emphasizing the construction of the optimal maximum likelihood (ML) processor. Simulations indicate a potential threefold reduction in error rates [6]. The article introduces an algorithmic program designed to address the limitations of Maximum Likelihood Sequence Detection (MLSD) in synthetic sequence systems by decoding spheres efficiently. The algorithm's complexity is analyzed, transcending heuristic methods and substantiated through simulation. The underlying challenge involves deciphering the real bases of a polymer sample from a given signal, posing a problem analogous to maximum likelihood sequence detection (MLSD). The article explores DNA ordering, most possibility sequence detection and bead decoding [7]. Sequencing deoxyribonucleic acid (DNA) is a pivotal element in contemporary research, opening avenues across diverse fields such as genetics, meta-genetics, and phylogenetics. This paper delves into the realm of machine learning in a scope where the instance space is polynomial in size. Traditionally, attention has been focused on learnability concerning thought categories with super-polynomial sample spaces, neglecting trivially polynomial learnable concepts [8]. A novel approach to DNA sequencing is presented in this paper, adopting a sequencing-via-synthesis methodology. The inherent nonidealities of this noisy switched linear gadget, parameterized by the unknown DNA sequence, are explored. The base calling challenge is framed as a boundary detection issue, aiming to identify device boundaries and minimize the risk of cryptographic errors [9].

An efficient method for DNA sequence Review on a diverse system, specifically the Xeon, is detailed in this paper [10]. Heterogeneous computing systems, incorporating CPUs and Xeon Phi devices, are discussed along with their background and everyday usage in regular expression matching. The experimental environment and results are presented, highlighting a comparison with contemporary related work [11]. Allele-specific expression (ASE), gauged through RNA sequencing, considers varying expression levels of different alleles [12]. Although ASE plays a role in genetic diseases by regulating penetrance or injury severity, its clinical application relies on DNA sequences, neglecting nuances like ASE. To harness the benefits of ASE without RNA sequencing, only predicting the forfeit of DNA mutations is necessary [13]. The application of ML to optimize the composition of particular sequencing panels emerges as a capable avenue for improving the detection of circulating tumour DNA (ctDNA) mutations in patients with this disease. In in-silico screening, Panel 2 outperformed alternatives in identifying ctDNA changes associated with cancer. One panel was derived from a grouping of current panels, while the previous one emphasized the frequency of neoplastic modifications. Methods like these are crucial for detecting alterations in cell-free DNA from heterogeneous cancer patients, achieving the necessary sensitivity for early detection at a reasonable cost [14]. In their "DNA act-Ran" proposal, the authors introduce a digital DNA ordering mechanism for ransomware finding using ML. DNA act-Ran leverages the principles of a digital DNA ordering scheme, and k-mer iterates the vectors for ransomware pathing tools. ML algorithms are useful to localize ransomware, and the feasibility and efficiency of the strategy are confirmed using a consistent dataset. Various benchmarks are employed to assess the proposed DNA act-Ran, showcasing its potential in ransomware detection [15].

This paper introduces BioSeq Analysis, an organic sequence analysis platform that facilitates machine learning techniques for process prediction. BioSeq Analysis continually evolves by integrating newly designed for the objective of intelligence, which is accomplished by utilizing advanced ML algorithms and various features systems in bioinformatics [16]. However, its application has been limited to sequence-level analysis tasks, necessitating an intelligent tool capable of generating diverse indicators for assembly- and sequence-level biological analysis. BioSeq Analysis 2.0 addresses this gap, standing as the first tool for generating indicators for assembly-level biological sequence analysis tasks. Experimental results indicate that the predictors developed by BioSeq Analysis 2.0 perform as well as, or even better than, general progressive predictors [17]. The findings of this paper suggest that artificial intelligence (AI) approaches offer a promising avenue for predicting drug resistance in infectious diseases. The exploration involves experimenting with machine learning techniques to predict deoxyribonucleic acid information based on drug resistance in Mycobacterium tuberculosis. The machine learning (ML) strategy demonstrated precise drug resistance prediction, achieving accuracy close to 100% and an area under the curve (AUC) of 1, indicating its reliability in tuberculosis prognosis. This result underscores the promising potential of AI approaches in tuberculosis drug resistance prediction. Additionally, the study highlights the data-specific nature of model performance, suggesting that parameter normalization can marginally enhance overall accuracy [18]. The proposed methodology deviates from several existing machine learning (ML) algorithms [15].

A significant challenge lies in characterizing the distinctive properties affected by specific diseases. Genomic research offers the ability to discern different protein features by categorizing DNA groups into existing classifications, thereby identifying and characterizing the associated genes. This paper analyses their corresponding biological applications and significance by exploring four common applications of machine learning in DNA sequencing data—DNA sequence alignment, classification, clustering, and pattern mining. It systematically summarizes recent research in DNA pattern mining by local and global analysts, addressing significant issues and proposing future research directions and models. Integrating more biological domains and artificial intelligence for user-friendly mining results is anticipated in future research [19]. This paper reviews gene classification methods using machine learning techniques, offering a brief overview of bioinformatics, literature, and key aspects in DNA sequencing through machine learning [20]. DNA (deoxyribonucleic acid) stands as a crucial macromolecule for all existing organisms, and DNA ordering, a technique for determining the sequence of nucleic acid bases, has become indispensable. The paper demonstrates the precise classification of double-stranded DNA molecules using artificial intelligence techniques applied to experimental quantum transport data. The accuracy exceeds 99.9% for primarily distinct molecules, allowing for the differentiation of even an individual nucleotide pair of E. coli DNA with 96% accuracy or higher [19]. In this study, CNN Bi-direction LSTM [21], CNN-LSTM [22], and CNN [23] formations were employed, utilizing tag and K-mer coding for DNA sequence classification. The system evaluation involved various classification methods, revealing that CNN and CNN bidirectional LSTMs using K-mer 20 encoding achieved accuracies of 93.1% and 93.1%, respectively, in the test data. The comparison of 3 DL techniques—CNN Bi-direction LSTM, CNNLSTM and CNN—highlighted the superiority of CNN encoding labels, albeit with lower test accuracy. K-mer encoding emerged as the optimal approach for achieving high test and validation accuracy. The evaluation of this dataset should consider multiple metrics, including precision, recall, sensitivity, and specificity [24].

10.3 INSPIRATION AND PURPOSE

Within the health sector, a wealth of data is available, offering the potential to predict an individual's future. Leveraging "machine learning and deep learning models," we aim to expedite DNA ordering and estimate processes. The goal is to provide medical practitioners with advanced technologies that facilitate the application of data to models for future predictions. Implementation hinges on the nature of the data employed and the specifics of the columns or features utilized in constructing the machine learning model.

Recognizing the time-intensive nature of the initial stages of this process, accuracy cannot be solely relied upon. Time is a critical component in the context of ML and DL models for integration into DNA study. The primary aim is that enhance prediction correctness within the prototype, unveiling valuable imminent and attaining the majority accurate principles for CNN, NB algorithms, decision trees, RF and Transform Learning. The overarching aim involves comparing the outcomes of ML algorithms with those of deep learning algorithms. Through this comparative analysis, the objective is to discern the superior performer among the five separate algorithms.

10.4 DNA ORDERING AND CATEGORIZATION

The process of categorization is employed to expect the group of substances with unidentified labels, utilizing information attained from a sample set. Particularly effective with distinct variable or definite data, classification proves to be an invaluable tool for the analysis of DNA data. In the realm of DNA sequencing, this technique is leveraged to scrutinize sequence similarities, be it in structure or function, and expect various series groups or modules, such as task and association. Categorization plays a vital role in identifying genes within a DNA particle.

CNN method offer a powerful method for categorizing DNA sequences, drawing inspiration from their successful applications in tasks like Email spam discovery and sentiment categorization in Grammarly. Furthermore, CNNs excel in extracting features from raw datasets. However, presenting unprocessed data as input to a CNN for feature retrieval and label prediction poses a challenge. The input has to undergo conversion into a mathematical representation.

To address this, regular text (phrases) can be encoded in the form of numerical by generating a key value where words serve as values and exact numbers act as inputs. However, DNA sequences differ from regular textual data as they don't consist of words; instead, they form a single 100-character word without spaces. To overcome this challenge, k-mer encoding is employed to transform the series into words, facilitating one-hot encoding based on the k-mer (typically of size six) key value to change it into a numerical format. This approach allows the utilization of any data categorization algorithm for DeoxyriboNucleic Acid classification.

Once the DNA order undergoes this numerical preprocessing, it becomes suitable for insertion into the Convolutional Neural Network replica for training. The Convolutional Neural Network structural design incorporates a convolutional layer responsible for extracting required features from the input, constructing a feature map. Additionally, a max-pooling layer reduces sizes, outputting only the main crucial data for further analysis.

10.5 PROPOSED SYSTEM

10.5.1 ARCHITECTURE

While the Future design may appear commonplace in its implementation, complexities arise in comprehending the dataset. The dataset comprises DNA sequences, posing a challenge due to its inherently cryptic nature. The presence of unwanted symbols necessitates a pre-processing methodology to enhance clarity and facilitate a more intuitive understanding. A detailed explanation of the proposed methodology's architecture is provided in Figure 10.1.

1. **Data Repository**: The data repository serves as the central storage facility for all information in its raw form. Raw data, essential for processing, is collected from diverse sources and housed in this repository, often referred to as a warehouse. This repository possesses the capability to store various types of data without any size restrictions.
2. **Data Preprocessing**: Data preprocessing involves the crucial step of cleaning the dataset. This process aims to eliminate any unwanted or irrelevant information present in the dataset. Machine learning models are not equipped to handle extraneous symbols or irrelevant information, making data preprocessing an essential step in preparing the data for analysis.
3. **Data Splitting**: The provided dataset undergoes a division into two subsets: the training set and the test set. Typically, researchers adopt one of three common compositions: 70%–30%, 75%–25%, or 80%–20%. In these configurations, the larger proportion is allocated to the training set, while the smaller Section is reserved for Validation purposes.
4. **Implementation of Prediction Models**: The execution of the prediction model involves the implementation of five distinct algorithms, technically referred to as models. These encompass DT, RF, NB, CNN, and TL.

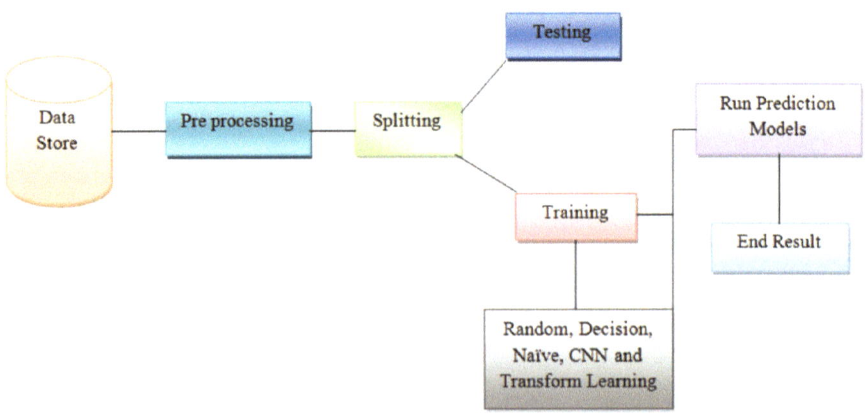

FIGURE 10.1 Proposed methodology for DNA sequence classification.

10.5.2 DEPLOYMENT OF ALGORITHMS IN THE PROPOSED SYSTEM

Each Machine Learning (ML) model outlined in the proposed system is discussed independently. The objective is to understand their functionality on the dataset and the techniques employed for problem resolution.

1. **Implementation of Decision Tree Algorithm**: The implementation of Deoxyribonucleic Acid (DNA) sequencing using the Decision Tree (DT) algorithm follows these steps:
    a. The DT algorithm utilizes the Gini Technique to compute the possibility of feature combinations.
    b. The Gini technique calculates both purity and impurity to assess potential losses in the dataset.
    c. The DT follows a top-down structure:
        i. The source node acts as the root.
        ii. Internal nodes contain intermediary data.
        iii. Leaf nodes store classification results.
    d. Classification occurs through a Yes/No clause traversal.
    e. Accuracy and recall are computed using a confusion matrix.
2. **Implementation of Random Forest Algorithm**: The Random Forest (RF) algorithm extends the decision tree by using multiple trees:
    a. A training set of K data points is selected.
    b. Decision trees are formed based on random subsets.
    c. The final classification is determined based on majority voting.
    d. The algorithm follows these steps:
        i. Select K data points for training.
        ii. Construct decision trees on subsets.
        iii. Choose N decision trees to form an ensemble.
        iv. Perform classification based on votes from N trees.
3. **Implementation of Naive Bayes Classifier**: The Naive Bayes classifier employs probability-based predictions:
    a. Features are selected for classification.
    b. Variables are converted into a mathematical format.
    c. The probability of each event occurrence is computed.
    d. The dataset is split into training and testing sets.
    e. The model is trained and tested for performance.
    f. Classification is based on computed probabilities.
4. **Implementation of Convolutional Neural Network (CNN)**: The CNN implementation follows these steps:
    a. Label encoding and one-hot encoding are applied to data.
    b. The training sample size K is determined.
    c. An embedding layer is created using K and encoded labels.
    d. A neural network (NN) is constructed with hidden layers.
    e. The CNN model is trained and tested iteratively.

5. **Deep Transform Learning Methodology**: This methodology utilizes deep learning networks for DNA sequencing:
   a. An embedding layer is created with different variables.
   b. Convolutional Neural Networks are trained on different subsets.
   c. Independent results are generated for each subset.
   d. The model integrates these results for final classification.
   e. New data is introduced, and steps are repeated iteratively.

10.6 IMPLEMENTATION

Our work revolves around the application of ML and DL methods to perform DNA sequencing. The human DNA, characterized by sequences such as ATGC, serves as the primary focus. Leveraging a dataset obtained from Kaggle, named "`human_data.txt`," we embarked on a classification algorithm endeavor. The objective is to classify DNA sequences into specific gene classes. Essential libraries such as pandas, NumPy and matplotlib have been imported for data manipulation and visualization. The dataset contains human DNA sequences and their respective classes. After processing this dataset, we now possess sequences paired with their corresponding classes. The goal is to predict the class based on the given sequence. This sequence may represent gene sequences or DNA sequences unique to individual human beings, categorized into various classes. In addition to the human dataset, we have obtained datasets for chimpanzees and dogs from Kaggle. These datasets follow a similar structure, comprising sequences and associated classes. The overarching aim is to apply classification algorithms to predict the class of a given sequence, extending our understanding of DNA sequencing through ML and DL methodologies and different classes are used to classify them. In addition to this specific dataset, we also obtained other datasets such as chimpanzee data and dog data from Kaggle. These datasets also contain sequences and classes.

10.7 MODEL ASSESSMENT AND ANALYSIS

This paper conducts a comprehensive comparison and analysis of the proposed system, evaluating its performance across various ML and DL Techniques to establish the model's accuracy. The models under consideration include Naive Bayes, Random Forest, CNN, Decision Tree, and Deep Transform Learning. The NB method exhibited superior correctness, achieving an impressive 98.70% in ML, while the deep transform learning algorithm demonstrated a commendable correctness of 95.37% in DL as shown in Table 10.1. The study highlights the importance of precision when classifying seven unique categories, revealing results for other models as well. For the Decision Tree model, an accuracy of 83.60% was attained, whereas the Random Forest model achieved a higher accuracy of 92.60%. The CNN model demonstrated a competitive accuracy of 91.65%. These findings collectively affirm the efficacy of the proposed model in DNA sequence categorization. Notably, the Naive Bayes model outperformed the others, achieving a remarkable accuracy of 98.70%.

TABLE 10.1
The Accuracy Metrics for All Algorithms Performance

Algorithm	Accuracy(%)
CNN	91.65
Random Forest	92.60
Naive Bayes	98.70
Decision Tree	83.60
Transform Learning	95.37

10.8 CONCLUSION

In conclusion, this study undertook a comparative analysis of five classification algorithms, Incorporating both conventional ML and DL methodologies. The algorithms investigated were DT, RF, NB, CNN, and deep Transfer Learning, employing k-mer and label Encoding techniques for DNA sequence classification. The paper extensively examined and compared various DNA sequence string encoding methods, including k-mer counting, label encoding and One-hot Encoding. Additionally, the study incorporated Natural Language Processing (NLP) bag-of-words techniques using count vectorizer for processing DNA sequence strings. The findings revealed that the Naive Bayes algorithm, when coupled with k-mer encoding, outperformed Decision Tree and Random Forest in terms of accuracy. Moreover, Transfer Learning with label encoding demonstrated superior accuracy compared to CNN with label encoding. Notably, Naive Bayes exhibited the highest accuracy of 98.70% among all algorithms, surpassing the proposed Transform Learning algorithm. Consequently, in comparison to label encoding, k-mer counting emerged as a more effective approach for our datasets. The evaluation of the dataset involved diverse metrics such as precision, recall, and accuracy.

REFERENCES

1. Babak Alipanahi, Andrew Delong, Matthew T Weirauch, and Brendan J Frey. Predicting the sequence specificities of DNA-and RNA-binding proteins by deep learning. *Nature Biotechnology*, 33(8):831–838, 2015.
2. Jian Zhou and Olga G Troyanskaya. Predicting effects of noncoding variants with deep learning-based sequence model. *Nature Methods*, 12(10):931–934, 2015.
3. Jian Zhou and Olga G Troyanskaya. Deep learning sequence-based ab initio prediction of variant effects on expression and disease risk. *Nature Genetics*, 50(8):1171–1179, 2018.
4. Seonwoo Min, Byunghan Lee, and Sungroh Yoon. Deep learning in bioinformatics. *Briefings in bioinformatics*, 18(5):851–869, 2017.

5. Christof Angermueller, Tanel Parnamaa, Leopold Parts, and Oliver Stegle. Deep learning for computational biology. *Molecular systems biology*, 12(7):878, 2016.
6. Stephen W Davies, Moshe Eizenman, and Subbarayan Pasupathy. Optimal structure for automatic processing of DNA sequences. *IEEE transactions on biomedical engineering*, 46(9):1044–1056, 1999.
7. Ting Wu and Haris Vikalo. Maximum likelihood DNA sequence detection via sphere decoding. In *2010 IEEE International Conference on Acoustics, Speech and Signal Processing*, pages 586–589. IEEE, 2010.
8. Ming Li. Towards a DNA sequencing theory (learning a string). In *Proceedings [1990] 31st Annual Symposium on Foundations of Computer Science*, pages 125–134. IEEE, 1990.
9. Helmy Eltoukhy and Abbas El Gamal. Modeling and base-calling for DNA sequencing-by-synthesis. In *2006 IEEE International Conference on Acoustics Speech and Signal Processing Proceedings*, volume 2, pages II–II. IEEE, 2006.
10. Suejb Memeti and Sabri Pllana. Accelerating DNA sequence analysis using intel (r) xeon phi (tm). In *2015 IEEE Trustcom/BigDataSE/ISPA*, volume 3, pages 222–227. IEEE, 2015.
11. Suejb Memeti and Sabri Pllana. A machine learning approach for accelerating DNA sequence analysis. *The International Journal of High Performance Computing Applications*, 32(3):363–379, 2018.
12. Matti Pirinen, Tuuli Lappalainen, Noah A Zaitlen, GTEx Consortium, Emmanouil T Dermitzakis, Peter Donnelly, Mark I McCarthy, and Manuel A Rivas. Assessing allele-specific expression across multiple tissues from RNA-seq read data. *Bioinformatics*, 31(15):2497–2504, 2015.
13. Zhenhua Zhang, Freerk van Dijk, Niek de Klein, Marielle E van Gijn, Lude H Franke, Richard J Sinke, Morris A Swertz, and K Joeri van der Velde. Feasibility of predicting allele specific expression from DNA sequencing using machine learning. *Scientific reports*, 11(1):10606, 2021.
14. Clinton L Cario, Emmalyn Chen, Lancelote Leong, Nima C Emami, Karen Lopez, Imelda Tenggara, Jeffry P Simko, Terence W Friedlander, Patricia S Li, Pamela L Paris, et al. A machine learning approach to optimizing cell-free DNA sequencing panels: with an application to prostate cancer. *BMC cancer*, 20:1–9, 2020.
15. Firoz Khan, Cornelius Ncube, Lakshmana Kumar Ramasamy, Seifedine Kadry, and Yunyoung Nam. A digital DNA sequencing engine for ransomware detection using machine learning. *IEEE Access*, 8:119710–119719, 2020.
16. Bin Liu. Bioseq-analysis: a platform for DNA, RNA and protein sequence analysis based on machine learning approaches. *Briefings in bioinformatics*, 20(4):1280–1294, 2019.
17. Bin Liu, Xin Gao, and Hanyu Zhang. Bioseq-analysis2. 0: an updated platform for analyzing DNA, RNA and protein sequences at sequence level and residue level based on machine learning approaches. *Nucleic acids research*, 47(20):e127–e127, 2019.
18. W Hadikurniawati, MT Anwar, D Marlina, and H Kusumo. Predicting tuberculosis drug resistance using machine learning based on DNA sequencing data. In *Journal of Physics: Conference Series*, volume 1869, page 012093. IOP Publishing, 2021.
19. Aimin Yang, Wei Zhang, Jiahao Wang, Ke Yang, Yang Han, and Limin Zhang. Review on the application of machine learning algorithms in the sequence data mining of DNA. *Frontiers in Bioengineering and Biotechnology*, 8:1032, 2020.

20. Yiren Wang, Mashari Alangari, Joshua Hihath, Arindam K Das, and MP Anantram. A machine learning approach for accurate and real-time DNA sequence identification. *BMC genomics*, 22:1–10, 2021.
21. Muhammet Fatih Aslan, Muhammed Fahri Unlersen, Kadir Sabanci, and Akif Durdu. Cnn-based transfer learning–bilstm network: A novel approach for covid-19 infection detection. *Applied Soft Computing*, 98:106912, 2021.
22. Beakcheol Jang, Myeonghwi Kim, Gaspard Harerimana, Sang-ug Kang, and Jong Wook Kim. Bi-lstm model to increase accuracy in text classification: Combining word2vec cnn and attention mechanism. *Applied Sciences*, 10(17):5841, 2020.
23. M Suriya, V Chandran, and MG Sumithra. Enhanced deep convolutional neural network for malarial parasite classification. *International Journal of Computers and Applications*, 44(12):1113–1122, 2022.
24. Hemalatha Gunasekaran, Krishnasamy Ramalakshmi, A Rex Macedo Arokiaraj, S Deepa Kanmani, Chandran Venkatesan, C Suresh Gnana Dhas, et al. Analysis of DNA sequence classification using cnn and hybrid models. *Computational and Mathematical Methods in Medicine*, 2021, 2021.

Index